Landscape-ecological Planning LANDEP

László Miklós · Anna Špinerová

Landscape-ecological Planning LANDEP

Springer

László Miklós
Faculty of Ecology and Environmental Sciences
Technical University in Zvolen
Zvolen, Slovakia

Anna Špinerová
Faculty of Ecology and Environmental Sciences
Technical University in Zvolen
Zvolen, Slovakia

ISBN 978-3-030-06774-8   ISBN 978-3-319-94021-2 (eBook)
https://doi.org/10.1007/978-3-319-94021-2

Softcover re-print of the Hardcover 1st edition 2019

This Springer imprint is published by the registered company Springer Nature Switzerland AG
The registered company address is: Gewerbestrasse 11, 6330 Cham, Switzerland

Preface

Landscape planning, landscaping, is a very attractive concept for geographers, ecologists, landscape ecologists, but also for architects, urban planners, environmentalists, conservationists and even the public. The word ‘landscape’ softens the technocratic perceptions of the word ‘planning’ on one hand, but on the other hand, it toughens the romantic views and brings contemplations over landscape to a different level. Thus, it hides many challenges and opportunities even for those, who do not entirely understand their essence. Indeed, the great boom of studies that discuss the topic of the landscape as well as the astounding number of authors—nature lovers—proves that it is the uncertain definition of the whole subject which offers the best representation of reality. When I say that it is probably the most fascinating field of all naturalists and non-naturalists whose profession has to do something with space, territory, region, landscape I may be expressing my subjective opinion, but doubt that I am the only one to hold it.

However, will landscape planning live up to the expectations of those who have thrown themselves into the arms of this scientific discipline, wishing to honour not only their own commitments, but also to be professionally committed or perhaps to live off it? I dare to say I doubt that many, if any, will achieve that. Simply put, there are more nature lovers than job opportunities available for landscape planners. There is nothing strange about this as there are not enough professional job vacancies even in the field of nature preservation to cater for all nature lovers and there are not enough chairs in orchestra to seat all the music lovers, and there are far fewer serious publishing houses who are willing to pay writers than there are literature lovers. Furthermore, I am concerned about the fate of all the results attained as a product of this ongoing enthusiasm. Are they ever going to be exploited, accepted, respected or will they reap hate from those affected by these results? This concern, however, is not new at all, as not each piece of work of a composer, writer or architect is always played, published or built. One needs to come to terms with this fact, realise how many of us, and how much effort needs to be exerted to spread education and awareness with regard to this fascinating area.

But what about the almost identical term—landscape-ecological planning? First, it needs to be stressed that this publication was originally written in the Slovak language, and so in other languages, other countries, the nomenclature might differ. In Slovakia, however, the laws define *landscape-ecological planning*, and even AGENDA 21 from the well-known conference on the environment in Rio de Janeiro in 1992 recommends the government to implement landscape-ecological planning (LANDEP). LANDEP—although addressing the same beautiful landscape as a nature lover or an artist, is a harder nut to crack. The landscape discussed in LANDEP does not require only a good appreciation of visible beauty, but also hidden, but exceptionally important features which can be examined only after serious study of, e.g. the geological base, soil, climate as well as other ongoing and certainly less pleasant phenomena such as erosion, or the daily activities of man—ploughing, farming and construction. The output is no less prosaic. The law stipulates that the results of LANDEP are the regulative for technical-planning processes, which means that they need to be based on solid arguments. The good news is that this is the right way in which the outputs of landscape-ecological planning can affect the development of the landscape in the long run. Even this gives rise to various questions and concerns. Do the institutional tools really function the way the law stipulates? Is the landscape-ecological plan for each survey for a land-use plan in compliance with the relevant law, is land consolidation performed to improve the ecological stability and are the water management effects of the territorial system of ecological stability incorporated into river basin management plans and integrated into landscape management in accordance with the law? The answers will hardly be affirmative for all three questions.

We scientists are well aware that the situation can be remedied by continuous improvement of the methods and methodologies through research of the landscape as well as the methods of the application of the findings into spatial planning processes.

This is the landscape-ecological planning our publication aims to discuss further.

Zvolen, Slovakia

László Miklós
Anna Špinerová

Acknowledgements

The publication is the result of the research supported by the grant agency KEGA Project No. 013TU Z-4/2016 and by grant agency VEGA Project No. 1/0096/1614-0735 and VEGA Project No. 2/0066/15.

Contents

Abbreviations

ABC	Abiotic Complex (Abiocomplex)
CLS	Current Landscape Structure
DTM	Digital Terrain Model
EC	European Commission
EIA	Environmental Impact Assessment
ELC	European Landscape Convention
EU	European Union
FAR	Folk Architecture Reservation
FMP	Forest Management Plan (Recently FCP—Forest Care Programme)
FMU	Forest Management Unit
GIS	Geographic Information Systems
HPZ/SZ	Hygienic Protection Zone/Safety (buffer) Zone
ILM	Integrated Landscape Management
INSPIRE	INfrastructure for SPatial InfoRmation in Europe
KEGA	Cultural and Educational Grant Agency of the Ministry of Education, Science, Research and Sport of Slovak Republic
KPP	Complex Agricultural Soil Survey of Slovakia
LANDEP	LANDscape Ecological Planning
LEB ILM	Landscape Ecological Basis of Integrated Landscape Management
LEC	Landscape-Ecological Complex
LTSES	Local Territorial System of Ecological Stability
m a.s.l.	Meters Above Sea Level
MAB	UNESCO's Man and the Biosphere Programme
MCR	Memorial City Reservation
MoE SR	Ministry of the Environment of the Slovak Republic
MPRV SR	Ministry of Agriculture and Rural Development of the Slovak Republic

NATURA 2000	Network of Sites of Community Importance and Protected Bird Areas
NFC	National Forest Centre
NFH	Non-forest Habitats
NISI	National Infrastructure of Spatial Information
NM	Nature Monument
NNM	National Nature Monument
NNR	National Nature Reservation
NP	National Park
NR	Nature Reservation
OECD	Organisation for Economic Co-operation and Development
PA	Protected Area
PBA	Protected Bird Area
PLA	Protected Landscape Area
PLS	Primary Landscape Structure
PT	Protected Tree
PWMA	Protected Water Management Area
PZ	Protection Zone
REPGES	REpresentative Potential GEoecoSystems
SAS	Slovak Academy of Sciences
SCI	Site of Community Importance
SEA	Strategic Environmental Assessment
SEC	Socio-Economic Complex
SEF	Socio-Economic Factor (Phenomenon)
SGIDŠ	State Geological Institute of Dionýz Štúr
SkEA	Slovak Environmental Agency
SLS	Secondary Landscape Structure
SNC SR	State Nature Conservation of the Slovak Republic
SR	Slovak Republic
SSCRI	Soil Science and Conservation Research Institute (Previously RISF)
STN	Slovak Technical Norm (Standard)
TLS	Tertiary Landscape Structure
TSES	Territorial System of Ecological Stability
UNESCO	United Nations Educational, Scientific and Cultural Organization
USDF	Unit of Spatial Division of Forest
WRI	Water Research Institute
ZB GIS	Fundamental Database (of topographic objects) for Geographic Information System

List of Figures

List of Tables

Introduction

It needs to be explained right at the beginning why we decided to devote the monograph to the landscape-ecological planning when there are a number of works discussing similar issues dealing with land, forests, waters. The major reason to present the essential features of the methodology of *landscape-ecological planning* (*LANDEP*) is a fact that the methodology proposed by the scientific institute described in this book is *reflected in the law and incorporated to the practice* of *particular spatial planning procedures in Slovakia*. Nevertheless, in spite that the frame and core of LANDEP method has been presented in many scientific forums as well as in scientific studies at home and abroad, since the methodology is relatively complex, in complete form has never been published in English. We dare to claim that even the translated publication cannot provide a fully comprehensive overview of the methodology, since the LANDEP has open character which can adopt new methods and, therefore, the partial steps are continually innovated. A quasi-complete overview could be theoretically provided only with a set of dozens of landscape-ecological plans, prepared with respect to legally adopted planning procedures, which is not possible to contain in a single methodological monograph, neither would have reason to collect them to whatever publication.

Landscape planning or relative activities are neither new nor unique. Nevertheless, we argue that the efforts of scientific teams have not entirely reached their goals, and only some of them have been really systematically accepted or found their way into social practice. Exactly, it is one of the notable—and probably rare—feature of LANDEP, which developed at a scientific institute—in the Institute of Landscape Ecology of the Slovak Academy of Sciences in Bratislava since 1967, was gradually applied to the projecting and planning practice of the 1980s, and later incorporated as legal norms of spatial planning of Slovakia between 1992 and 2000 (Ružička and Miklós 1979, 1982a, 1990).

Until 1990, the LANDEP was incorporated to spatial planning practice only because of the good judgement of the leaders of planning and projecting companies and organisations such as Stavoprojekt in Žilina and Banská Bystrica, URBION in Bratislava, Office of the Chief Architect in Bratislava, the Agricultural Project Institute in Bratislava. After the political changes in 1989, the LANDEP method together with the almost simultaneously developed method of the Territorial System of Ecological Stability *TSES*—which is a specific eco-network projecting method in Slovak and Czech Republics—were incorporated into the legal system of spatial planning, namely into the amendments of Act 50/1976 on Territorial Planning and Building Order as amended by Act 262/1992 and, in particular, by Act 237/2000. Throughout this period, both methods became regular and routine procedures in the practical processes in the Slovak Republic.

We developed, applied and we still consider the LANDEP method a base for the *ecologically optimal* spatial organisation and utilisation of landscape, as such as a base for the *integrated sustainable landscape management*, and, consequently a base for the spatial/complex protection of the environment (in contrast to the technological protection of individual components of the environment).

It is to be underlined at the very beginning that the LANDEP method differs from those landscape plannings which are based on visual structures of landscape, or are of a character of landscape architecture, greenery design, garden and park projecting. In LANDEP, the accent is given to the functionality of geosystems for activities of the man, very much depending on the *vertical bonds* of the elements of geosystems from geological substratum up to atmosphere. Mentioned Acts stipulate that the results of LANDEP are the *regulative for technical-planning processes*. This means that they need to be based on solid arguments.Therefore, it was in all period based on the landscape ecological science, pillared by the understanding of the landscape as the geosystem. The TSES—as a specific branch and complementary method to the LANDEP consequently focuses on the *spatial structure* of the elements of the landscape. Thus, the mentioned two procedures complete both basic requirements on the optimum use of the landscape—the vertical–functional and the spatial–organisational optimisation.

The LANDEP methodological approach, from its first adoption in 1979 has undergone many improvements. So as various experts became involved in the development process of the methodology, they improved particularly the partial methods and procedures within each step, especially in the process of interpretations and evaluations, new techniques were adopted, as GIS. At the same time, it should be stressed that the methodological basis, material and temporal sequence of steps and logic of confrontation of landscape with the requirements of human activities is virtually unchanged.

The principal status of five steps of the LANDEP is still valid. The first two steps—*analyses and syntheses*—have the form of *fundamental research* and result in the design and characteristics of complex landscape-ecological-spatial units. The final two steps—*evaluations and proposals*—address the needs of *planning practice*. The intermediate third step of the methodology—*interpretations*—has the character of *applied research* and forms the arguments and criteria for the assessment of landscape for its utilisation by man (Ružička and Miklós 1982b; Miklós and Špinerová 2011).

We would like to note here that the LANDEP methodology was also recommended as an integrated approach to landscape management and natural resources in Agenda 21 from Rio Summit ´92 (Chapter 10, AGENDA 21).

Agenda 21, Chapter 10:
"Government on the appropriate level ... should:
Adopt planning and management systems that facilitate the integration of environmental components such as air, water, land and other natural resources, using ***landscape ecological planning (LANDEP)*** *or other approaches that focus on, for example an ecosystem or a watershed.*"

The presented publication aims to give a comprehensive description of the scientific and methodological surroundings of the methodic of *LANDEP*, as well as to the introduction of the methodical procedure of the individual methodical steps of LANDEP. We consider such a summarising work as important, since in the last period a huge amount of landscape ecological works named as landscape planning appeared with a very different content, built up on a very different level of the scientific approach, sometimes dealing with real planning just by name.

The Graphics and Tables

The graphics, maps, tables and matrixes are of utmost importance in the LANDEP method. Therefore, the publication is completed by a number of graphics and tables aiming to complete the picture on the method. They are of different characters. Partly, they amend the *methodical* parts of the book complementing the understanding of single methodical steps. In the chapters on evaluation and proposition the tables have even the core importance, the evaluation in LANDEP is based on the decision table. The other part of graphics and tables have *illustrative function*, they should document that the methodical steps are feasible, they have real implementation and were applied in real studies in model territories. So, they have a character of concrete examples which illustrate the methodical steps. Since they relate to one concrete model territory, they do not have real factual information relevance. It is to be also underlined that we included to the publication map illustrations from different territories, from studies in different time periods, in different scales. Some of the older figures which we consider from the methodical point of view as important are even hand drawings. In opposite, a number of illustrations is the result of computer processing with GIS techniques. These information were collected mostly in detailed scales, are basically stored electronically. Nevertheless, the hard copies presented in the publication cannot display all those details, they are obviously essentially generalised and on lower graphical quality. The reason we annexed them to the book is the illustration of methodical step. For the readers may be of higher worth not the map design but probably the map legend, which bears the information on the map content. We consider all figures fulfil the above-mentioned functions.

All figures, graphics and tables are created by the authors of the publication (including also copyright of the authors). The most comprehensive works were presented in the publications Miklós (1978)—model territory Gemerské Turce catchment, Miklós et al. (1985)—the Ecological General Model of Czechoslovakia, Miklós et al. (1986)—East Slovakian Lowland, Miklós and Hrnčiarová (2002)—Landscape Atlas of Slovak Republic, Miklós et al. (2011)—Ipeľ river catchment, Špinerová (2010, 2015)—Ilijský creek catchment.

References

Miklós L (1978) Náčrt biologického plánu povodia Gemerských Turcov. VEDA, Bratislava, Questiones Geobiologicae, 21:127

Miklós L et al (1985) Ekologický generel ČSSR. Časť SSR. I. etapa: Priestorová diferenciácia územia z ekologického hľadiska. Záverečná správa P16-121-402/01. ČSŽP Bratislava. ÚEBE CBEV SAV Bratislava, Stavoprojekt Banská Bystrica

Miklós L, Hrnčiarová T (eds) (2002) Atlas krajiny Slovenskej republiky. 1. vydanie, MŽP SR Bratislava, SAŽP Banská Bystrica

Miklós L, Izakovičová Z, Kanka R, Ivanič B, Kočický D, Špinerová A, David S, Piscová V, Štefunková D, Oszlányi J, Ábrahámová A (2011) Geografický informačný systém povodia Ipľa: Katalóg GIS a výber máp. Bratislava: Ústav krajinnej ekológie SAV: Katedra UNESCO, Fakulta ekológie a environmentalistiky, Technická univerzita vo Zvolene: Esprit Banská Štiavnica

Miklós L, Kozová M, Ružička M et al (1986) Ekologický plán využívania Východoslovenskej nížiny v mierke 1:25 000. Ekologická optimalizácia využívania VSN. ÚEBE SAV Bratislava, Slovosivo. III. diel, pp 5–312

Miklós L, Špinerová A (2011) Krajinno-ekologické plánovanie LANDEP [The Landscape-ecological Planning LANDEP]. Harmanec, VKU

Ružička M, Miklós L (1979) Teoretické a metodologické základy biologického plánovania krajiny. Záverecná správa úlohy VI-3-5/1. Bratislava, ÚEBE SAV

Ružička M, Miklós L (1982a) Landscape-Ecological Planning (LANDEP) in the Process of Territorial Planning. Ekológia (ČSSR), 1:297–312

Ružička M, Miklós L (1982b) Metodické poznatky ekologického hodnotenia územia pre zónu a sídelný útvar (na príklade Rimavskej Soboty). VEDA, Bratislava, Acta Ecologica, 9, 26, pp 74

Ružička M, Miklós L (1990) Basic premises and methods in landscape-ecological planning and optimisation. In: Zonnenveld IS, Forman RTT (eds) Changing Landscapes: an ecological perspectives, Springer Verlag, New York, pp 233–260

Špinerová A (2010) Krajinno-ekologické limity poľnohospodárskeho využitia Ilíjskeho potoka. VKÚ, a.s., Harmanec

Špinerová A (2015) Štruktúra krajiny ako regulátor dynamiky pohybu vody a materiálu. Zvolen, Vydavateľstvo TU vo Zvolene

Chapter 1
Principles, Theoretical and Methodological Background of Landscape-ecological Planning

Abstract The methodology of **landscape-ecological planning LANDEP** is reflected in the law and is incorporated to the practice of particular spatial planning procedures in Slovakia. First chapter of the book deals with different understandings of spatial planning procedures and compares them to LANDEP. LANDEP represents a spatial approach to the protection of the environment. Its methodology, content and practice correspond with the principles of the integrated landscape management defined since AGENDA 21 from Rio Summit 1992 in all relevant environmental political documents, in particular, the following:

- cross-departmental conception: sectorial integration, **partnership**,
- over-regionality: spatial integration, **regional cooperation**,
- complex approach to the environment: professional integration, **landscape-ecological complexity**.

The chapter defines the main institutional spatial planning procedures supported by relevant law in Slovakia and their relations to the integrated landscape management and LANDEP. These tools are divided into four groups as follows:

- the information base of integrated landscape management,
- landscape-ecological framework for integrated landscape management: the LANDEP, Territorial System of Ecological Stability TSES and the territorial planning,
- the sectoral spatial management and planning instruments as nature conservation and landscape protection; river basin management; flood prevention, agricultural land consolidations projects; forest management plans,
- instruments for control and assessment of the impact of planning procedures on the environment as environment impact assessment and strategic impact assessment (E.I.A., S.E.A) and Integrated Prevention and Pollution Control (IPPC).

As the LANDEP is by legislation assigned in Slovakia to the territorial planning, which according to its content represents the most integrating concept with respect to the aforementioned principles, the chapter discusses the act on territorial planning

L. Miklós and A. Špinerová, *Landscape-ecological Planning LANDEP*,
https://doi.org/10.1007/978-3-319-94021-2_1

in detail. The material object of LANDEP is defined as landscape as a geosystem. The final subchapter deals with those aspects of geosystems which relate to the particular methodical steps of the LANDEP procedure defines the relevant models, structures, elements and properties of geosystems including the most frequent values of indicators of properties.

Keywords LANDEP · Territorial planning · Integrated management · Planning tools · Geosystem

The LANDscape-Ecological Planning **LANDEP** (in Russian LANDshaftno-Ekologicheskaya Planirovka—hence the abbreviation responding to two languages, Ružička and Miklós 1982b, 1990) is primarily a kind of spatial planning—the planning of the ecological optimum **spatial organisation** and **functional utilisation** of the territory. It is to state that the definition and understanding of the terms relative to the landscape-ecological planning are far not unified (Finka et al. 2008). The following section explains briefly the relations of these terms.

1.1 What Is Landscape-ecological Planning?

The main planning aspects—the spatial organisation, arrangement of objects and activities, a form of space management—appear in all definitions of the most general term '***spatial planning***'. The Compendium of European planning systems defines the general task of all spatial plannings as '*the future distribution of activities in space*' (Compendium 1997). So, all spatial plannings should bear the aspects of **proposal, future, space and arrangement** (Europäische … 1983; Resolution … 1990; Williams 1996; European Spatial … 1999; Spatial planning … 2008). Koresawa and Konvitz (2001) define the role of spatial planning as '*coordination of sectoral policies such as transport, agriculture and environment*'. The spatial planning includes also the landscape-ecological elements, it concerns the whole territory and the total landscape (Fabos 1979; Spitzer 1995).

Nevertheless, many times the 'spatial' plans are driven and covers the planning of urbanised territories with less regards to rest of landscape (McHarg 1969; Golley, Bellot 1991; Hall 1992; Newman and Thornley 1996; Belčáková et al. 2003; Antrop 2004; Izakovičová et al. 2017). These approaches are close to plannings clearly oriented to urbanised territories, in spite of that the definitions intend to define them more generally, e.g. Cullingworth and Nadin (2006) define the role of '***town and country planning***' as '*coordination or integration of the spatial dimension of sectoral policies through a territorially-based strategy*'. So, the terms '***land-use planning, regional planning, urban planning, and urban design***' are often used interchangeably (Alterman 2001).

In concern to the presented book it is to be clarified that the spatial planning in Slovakia is marked in exact translation as '***territorial planning***' (územné plánovanie). Also, the very basic, still valid, legal tool is named as **Act on the Territorial Planning and Building Order** (Act No. 50/1976 Coll. and its later ammendments). So, we shall use the term 'territorial planning' throughout the book in the sense of spatial planning in all cases when it concerns the Slovakian practice (Hreško et al. 2003).

Another very frequently used term, which is by the content very similar to LANDEP, is marked as '***land-use plan***'. In slight contrary to the general spatial plans, the land-use plans tend to concentrate on non-urbanised part of the landscape, mostly to agricultural land. The compendium Spatial Planning (2008) defines the land-use planning as '*… systematic assessment of land and water potential, alternative patterns of land use and other physical, social and economic conditions, for the purpose of selecting and adopting the land-use options which are most beneficial to land users without degrading the resources or the environment …*'

The Canadian Institute of Planners (2013) included to the definition also the aesthetic aspect as '*land-use planning means the scientific, aesthetic, and orderly disposition of land, resources, facilities and services with a view to securing the physical, economic and social efficiency, health and well-being of urban and rural communities*'. This concept is covered by very rich literature of both theoretical and practical character (e.g. Young 1986, 1993; Land-use planning applications 1991; Guidelines 1993; Verhaeye 2004; Wehrmann 2012; Mederly et al. 2012; Rao et al. 2015).

The land-use planning in last decades is many times marked as **multifunctional planning**, considering the landscape a multifunctional entity (Brandt 2003; Brandt et al. 2000; Brandt and Vejre 2004; Mander et al. 2006).

To spatial planning procedures belong also tools for the environmental care in spatial sense, sometimes referred to as '***environmental planning***' (Buchwald 1980; Barančoková et al. 2010). In fact, all the above-mentioned spatial planning processes consider itself as important instruments for the environmental care and for sustainable development (Miklós 1992; Mederly and Hudeková 2005; Petrovič et al. 2011; Silberstein and Maser 2013; Brandt et al. 2013a, b).

Probably the most popular kind of spatial planning in last decades is the '***landscape planning***' (different to landscape-ecological planning!), which is oriented to the evaluation and arrangement of land cover, very much insisting on the values, aesthetics, cultural features, historical landscapes (Hreško et al. 2015). This approach is closed to '***landscape architecture***', sometimes marked as '***landscaping***'. The landscape planning, landscaping, has also very rich literature (Marsh 1997; Rega 2014; Ružička et al. 2010; Belčáková 2013a). However, numerous landscape planning projects were developed by landscape ecologists or geographers (Breuste et al. 2009; Kozová, Pauditšová, Mišíková 2010; Jones et al. 2013). This approach is very much supported by the provisions of the European Landscape Convention (2000).

Also, very popular among landscape ecologists is the projection of '***ecological networks***' (Fabos 1996; Ahrend et al. 1992; Cook and van Lier 1994; Jongman 1996). A specific method of ecological network planning in former Czechoslovakia has been developed parallel to the LANDEP method named "***Territorial System of Ecological***

Stability TSES" (Lőw 1984; Miklós et al. 1985; Buček et al. 1986; Jančura et al. 1994; Izakovičová 1996; Diviaková 2010). The main task of TSES project is the design of the spatial frame of TSES—the spatial distribution of biocorridors, biocentres and interactive elements—and the proposal of ecostabilising measures, which enables the functioning of this frame (Miklós, Diviaková, Izakovičová 2011).

We would like to point out in relation to above-defined spatial planning procedures that many concepts referred to as 'planning' and 'landscape planning' have the character of various applied studies, which may indicate the possibilities to apply their findings to planning practice, but they have **not reached the step of real planning** yet. The predecessors of LANDEP, so-called biological plans, were also of such a character (Drdoš 1967; Miklós 1978; Oťaheľ 1978). Other concepts, which aspired to be landscape plans, lacked the character of plans too, e.g. the concept of landscape synthesis and diagnosis (Drdoš et al. 1980; Mazúr et al. 1983). Although their name accurately reflects their content—they indeed contained syntheses which resulted in landscape diagnosis—they were not plans. Many other studies have a similar character, e.g. the studies on topochores (Hynek and Trnka 1981), highly elaborated concept of landscape potentials (Haase 1978; Drdoš 1983; Drdoš 1992; Tremboš 1993), and various other procedures of landscape 'evaluations'. They saw the most rapid development in the early 1990s (e.g. Zonneveld and Forman 1990; Reitsmaa 1990; Haase 1991; Barsch and Saupe 1993; Bastian and Schreiber 1994; Langevelde 1994; Žigrai 1994; Minár et al. 1995; Oťaheľ, Lehotský, Ira 1997). On the other hand, in the past, as well as, nowadays, there are landscape-ecological studies which, in spite that they are not named 'planning', show planning aspects, e.g. the 'classical' works of McHarg (1969), who discussed design with nature, but also other authors (Fabos 1979; Huba 1982; Buček et al. 1986; Haber 1990; Oťaheľ 1994; Halada et al. 1995; Hrnčiarová et al. 1997; Bezák et al. 2017). Perhaps, it was the boom of applied landscape-ecological works of the 1990s and their comparison with the real state of landscape-ecological planning which inspired J. Drdoš to conduct a realistic and critical evaluation of the activities of geographers and landscape ecologists in this area (Drdoš 1995).

In any case, since then no significant methodological breakthrough or quantitative advancement in development of the transformation of collected landscape-ecological results into planning practice has been observed (Kerényi 2007). It can be even argued that the applied part of the methodology—the methods of objective evaluation of the suitability of landscape properties for performing various social activities have not progressed sufficiently. We have rather seen a quantitative growth of studies discussing applied research. Numerous studies labelled as landscape planning represent a collection of application methods of landscape evaluation (Kozová and Bedrna 2003; Sklenička 2003; Kozová et al. 2007; 2010; Izakovičová and Moyzeová 2007; Moyzeová 2008; Breuste et al. 2009; Machar and Kovář 2010; Mizgajski and Markuszewska 2010; Izakovičová et al. 2011a). The application of GIS technologies has also greatly expanded the possibilities for evaluating spatial relations in the landscape (Miklós et al. 1986; Hreško et al. 2003; Gercsák 2011; Konečný et al. 2012; Kocsis et al. 2016a, b). Many other methods of analyses, syntheses and interpretations were continuously developed, even not always under the designation landscape ecology (Zonneveld 1995). According to Antrop (2013) '*more and more scientific disciplines borrow methods and techniques from others, especially when they offer innovations*'. This is one of the possible reason why there is a gap, although, hopefully, not a void anymore, between science and practice still persists (Antrop 2013).

Nevertheless, although these works do not have a planning character, it does not mean that they are not valuable scientific works. In opposite, they could produce important methodical knowledge usable also in planning procedures.

Recently, new applied landscape-ecological concepts have emerged which focus also on determination of value of the landscape for various human activities. This concept—slightly biased and influenced by fashion mainstream trends—is referred to as an evaluation of **ecosystem services** (Constanza 1997; Kienast et al. 2007; de Groot et al. 2010; Iverson et al. 2014; Grunewald and Bastian 2015). This concept penetrated also to the decision-making processes and to landscape management (Mederly et al. 2015). Although there is an indisputable positive influence of this concept on the overall acceptance of natural sciences in the development, it also raises a lot of, yet, unresolved methodological issues, scientific distortions and simplifications, whereas the potential of previous similar methodologies has certainly not been fully exploited (Bezák et al. 2017).

Most of the authors consider the landscape-ecology as a basic science for the above mentioned planning tools (Opdam et al. 2002; Izakovičová and Moyzeová 2007; Csima and Dublinszki-Boda 2008; Brandt et al. 2013). The landscape-ecological character of LANDEP stems from the Central European understanding of landscape ecology as a complex scientific discipline which studies the landscape as a geosystem with both its natural and socio-economic spheres (Armand 1949; Mičian 1982; Csorba 1987; Ružička and Miklós 1990; Haase 1991; Barsch et al. 1993; Tremboš 1993; Bastian and Schreiber 1994; Brandt 1999; Mizgajski and Markuszewska 2010). Thus, the object of landscape-ecological planning is **landscape as a geosystem** with its entire complexity of vertical and topical relations, and not only its visible components, as the term landscape is often assumed.

For the scientific basement of the landscaping could be considered the landscape ecology in the sense of Formann and Godron (1981) based on the assessment of spatial structure, sometimes other approaches. Many such landscape-ecological projects were developed '*not always under the designation 'landscape ecology', but as part of landscape stewardship, landscape architecture and, first and foremost, environmental or urban and landscape planning.*' (Zonneveld 1995).

Finally, to the family of ecologically biased spatial plannings belongs also our Landscape-Ecological Planning LANDEP.

Our definition of LANDEP:
LANDEP is a geosystem based purpose-oriented complex of ***applied*** landscape-ecological methods aimed at the ecologically optimum spatial organisation, utilisation and protection of landscape which results to the proposal of most suitable ***localization*** of demanded human activities within the given territory (what and where?) and successively to the proposal of ***necessary measurements*** ensuring the ecologically proper functioning of those activities on the given locality (how?) (Ružička and Miklós 1982b, 1990; Miklós and Špinerová 2011).

Some basic characteristics of the landscape-ecological planning LANDEP.

The LANDEP method is similar to many of the above-mentioned planning procedures, it differs from them in several important aspects, which are given as follows:

- the basic object of LANDEP is the landscape as a geosystem (explanations see in next chapters);
- LANDEP is focused on the whole-landscape space (to forests, agricultural land and urbanised landscape, too);
- LANDEP intends to evaluate the territory for all requested activities (rural, urban, industrial and agricultural);
- LANDEP is developed as a regular obligatory base for the legally defined territorial planning. In this planning, the properties of the landscape play role of the **obligatory regulatives**.
- Thus, the LANDEP has a routine application in the real planning practice, supported by law. This kind of spatial plan is quite rear (Hreško et al. 2008; Izakovičová and Moyzeová 2007; Szabó et al. 2012; Belčáková 2013b).

All important aspects of the plannings—namely the **proposal, future, space, land and arrangement**—have been reflected in LANDEP since its inception (Ružička and Miklós 1979). Since the output of LANDEP is a plan—the proposal of applicable landscape-ecologically optimum organisation and utilisation of a territory—in the form of delimited areas for certain activities, a proposal for landscape greenery, up to functional typisation and regionalisation of the territory—LANDEP is legally incorporated as a **basis and regulation** for spatial planning procedures, territorial plans, land-use plans and other sectoral plans. If various studies on landscape do not possess the aspect, they cannot be considered as planning.

As we mentioned in the introduction, the five subsequent steps of LANDEP—**analyses, syntheses** (these first two steps are of fundamental research character), **interpretations** (this is the third step of applied research character), **evaluations** and **proposition** (these are the last two steps of planning character) have not yet been surpassed. The logic is still valid no matter how many substages other methodologies have (compare with, e.g. 14 steps of the landscape plan proposed by Huba (1982); or ‘only’ two steps of synthesis and diagnosis in the concept of landscape syntheses by Mazúr et al. (1983).

In terms of the actual implementation of landscape-ecological planning, LANDEP, the situation can be considered slightly schizophrenic. Until 1989, neither LANDEP nor TSES were included to the legal spatial planning tools. Nevertheless, they were respected by decisive planning and projecting institutions as Stavoprojekt in towns Žilina and Banská Bystrica, URBION in Bratislava, Office of the Chief Architect in Bratislava, the Agricultural Project Institute in Bratislava. At the Stavoprojekt Banská Bystrica, a specialised atelier was created devoted to landscape-ecological planning, so, actually, all territorial plans were based on LANDEP method. Since 1990, both methods have been systematically introduced into the legislation, especially by "greening" the Act 50/1976 on territorial planning and building order in 2000 by the amendment 237/2000, and also into other spatial planning tools as the agricultural planning, water planning (see later). Thus, we can boldly declare that landscape-ecological methodologies are sufficiently implemented to the legal tools in the Slovak Republic. Nevertheless, a very serious problem of practical implementation of LANDEP as well as the TSES after this successful implementation to the law was that the decisive implementing tools of both—the legally defined territorial planning—saw a significant decline. The territorial planning, which enacts LANDEP and TSES, ceased to be a part of the Ministry of Environment of the Slovak Republic after 2002, its content fell under other two ministries, and finally ended up in the Ministry of Transport, Construction and Regional Development. However, it was not developed along this 'journey' and almost disappeared from the consciousness of landscape ecology, but what is worse, also actual practice. Territorial plans are being most often replaced by so-called 'zoning decisions', which cannot deliver on the application of ecological aspects.

The second major large-scale planning tool—projecting of agricultural land consolidations and arrangement—which should make provisions for TSES as stipulated by the law, are not projected enough and attention is often paid more to resolution and consolidation of land ownership relations rather than the TSES (Izakovičová et al. 2010). Consequently, TSES projects are produced often as separate projects, what is less systematic solution than the case when is projected in bond with territorial or agricultural planning.

There was some hope raised by the arrival of the new law on flood protection (Act 7/2010 Coll. on Flood Prevention), which clearly defined landscape-ecological aspects as a tool of the integrated watershed and river basin management.

In this regard, the relation between LANDEP and TSES should be mentioned. Since both methodologies were devised at the same institute, the Institute of Landscape Ecology, they are derived from the same landscape-ecological theoretical basis. Landscape-ecological analyses and syntheses constitute the basis for both LANDEP and TSES. TSES can be regarded as an **organic part of the proposal of ecologically optimal organisation and utilisation of the landscape**, and thus a part of LANDEP. In fact, LANDEP cannot be projected without respecting the elements of TSES as a part of optimal organisation, and also TSES cannot be established without being a part of an ecologically optimal organisation. If TSES is projected as a part of a LANDEP project—which is an ideal scenario—such complex project comprises both the following aspects:

- the **functional determination** of ecologically optimal use of the landscape and the ecostabilising measures based on vertical bonds of geosystem elements, as well as
- **spatial distribution** of the ecological network elements—the biocentres, bicorriors and interactive elements.

The projecting single TSES projects lies in more extensive assessment of the biotic component of landscape and the outputs contain more detailed characteristics of biocentres and biocorridos, but, in such cases, the projects often lack the ecostabilising measures.

The TSES is discussed in detail in many other publications (Lőw et al. 1984, 1995; Buček et al. 1986; Jančura et al. 1994; Izakovičová 1996; Diviaková 2010; Miklós et al. 2011).

1.2 Ecologisation of Landscape Management

1.2.1 Integrated Conceptions to the Environmental Care

In the second half of the last century, as a reaction to various environmental disasters, society began to tackle the issue of environmental protection as a necessary condition to preserve humanity and progress. To this day, environmental protection has expanded dramatically and can be divided into the following three fundamental streams:

- ❖ **technological** protection—today can be called 'classical' protection—protection of the individual **components** of the environment. This stream includes well-known and developed technologies of protection of water, air, soil, geological resources and waste management.
- ❖ **technical—sanitary** protection from the adverse environmental factors such as noise, radiation, vibration, heat and light pollution, allergens and aesthetic problems.
- ❖ **spatial-organisational** protection of the environment, which may be, nowadays, referred to as **integrated care on the environment/landscape as a whole**. This protection includes the maintaining the ecological functionality, biological balance, stability in the landscape, and biological diversity.

The current trend of sustainable development and the protection and exploitation of natural resources were according to the conclusions of the UN Conference on Environment and Development—the Rio Summit in 1992, World Summit on Sustainable Development (WSSD, Rio + 10 –) in Johannesburg in 2002, Conference on Sustainable Development (Rio + 20) in Rio the Janeiro in 2012, the UN Sustainable Development Summit 2015 in New York and other related conferences UNEP and CSD (Commission for Sustainable Development), UN/ECE, Council of Europe,

OECD, and the continuing evolution of environmental legislation in the EU, bear the spirit of the general principles of integration, in particular, the following:

- cross-departmental conception: sectorial integration, **partnership**,
- over-regionality: spatial integration, **regional cooperation**,
- complex approach to the environment: professional integration, **landscape-ecological complexity**.

These fundamental aspects have been gradually formulated into international policies and conventions. As the base, we can consider, the still valid and not yet overcome, Chapters of Agenda 21, particularly the Chapter 10 entitled 'Integrated approach to planning and management of land resources', as well as other chapters, such as chapter 13 on fragile ecosystems: sustainable mountain development, chapter 14 on sustainable agriculture and rural development, chapter 15 on conservation of biological diversity, chapter 17. on human settlements and others. Individual governments need to observe the principles when devising their methodical, legislative and economic instruments.

The first environmental politics in Slovakia after the political changes in 1989 named as the 'Ecologisation of the development of the society' was of clearly integrated character. It had four main principles, as follows:

- ecologisation of the awareness of the society,
- ecologisation of the legal and economic instruments,
- **ecologisation of the spatial organisation and utilisation of the landscape** and
- ecologisation of the technological processes (Miklós 1991, 2011a, b).

We considered that time the concept 'ecologisation' actually equal to the concept 'sustainable development'. It may be noted that it was two years before the Rio Summit 92. Later, the concept of the sustainable development has been broadly accepted, the world 'ecologisation' loos its popularity. Nevertheless, during last decades, also the concept of sustainable development underwent several difficult phases, even crisis, e.g. during the WSSD in Johannesburg in 2002 (Johannesburg 2002), or unsuccess of the Millennium Development Goals (United Nations 2000). These problems are caused mainly because of the very broad scope of the concept, non-concrete content, thousands of interpretations, very diverse use and misuse of the term. Since Rio Summit '92 several new approaches tried to substitute or support the sustainable development (Nesshöver et al. 2017), up to new UN Sustainable Development Goals (Transforming our World: The 2030 Agenda for Sustainable Development 2015), but also criticise it (Spangenberg 2016).

On other side, the concept 'ecologisation' is more addressing, clearly relates to the environment, expresses the effort to turn the humans towards environmental friendly activities. It is defined, e.g. as a civilizing process of the current political-economic régime, as a continuation of the economisation, as the ecologisation of the economy (Schmidt 1993), or as "the process of growing integration of environmental concerns (referred to as 'ecologisation') in European agricultural policies" (Deverre and De Sainte Marie 2008). The word ecologisation was again risen to the politics, e.g. on the UNECE Ministerial Conference 'Environment for Europe', 2011 in Astana, on the UN Conference on Sustainable Development in Rio de Janeiro, 2012—Rio +

20 (The Future … 2012), as well as in very diverse documents (Implementation … 2009; Lamine 2011; Tsybuliak 2015).

Among other important tools which implicit the ecologisation as integrating principles we can mention the following:

- international conventions, for example, the UN Convention on Biological Diversity, the Framework UN Convention on Climate Change and its protocols, the European Landscape Convention, Carpathian Convention (UNEP ROE), the EU Strategy for the Danube Region;
- Europe-wide initiatives and legislation, especially
  - concepts of Pan-European Ecological Network (The Pan-European … 1995) and the National Ecological Network (in Slovakia—territorial systems of ecological stability TSES);
  - NATURA 2000 and practically, entire modern nature conservation in Europe;
  - The Water Framework Directive (European Parliament and Council 2000/60/EC);
- national concepts of landscape or landscape-ecological planning, separate procedures or parts of other space-planning processes;
- the concept of integrated landscape management in national legislation systems;
- the Millennium Goals (UN Millennium Declaration 2000), in particular, the goal 7 (To ensure environmental sustainability) and 8 (To develop a global partnership for development), and the newest;
- Sustainable Development Goals (Agenda 2030) (in particular goals 6, 7, 13, 14, 15) (Transforming our World: The 2030 Agenda for Sustainable Development 2015).

All those integrative conceptions implicit both the
ecological and spatial
aspects as basic principles

The basic principles of **integrated** environmental care undoubtedly include **ecologically optimal spatial organisation and land use**, or its specific tools—ecologically optimised spatial planning processes. They are considered integrated tools for the sustainable development in the sense of aforementioned theoretical propositions: they organise human activities in the landscape as a **whole** (in the geosystem—see the following chapter), and this whole-landscape space integrates **all human activities** (Izakovičová et al 1997).

1.2.2 Spatial Approach to the Environmental Care

Although the development of a theoretical–methodological basis of the integrated approach to the management of the environment has been initiated, current practice is still dictated especially by the analytical, component approach. It focuses on the protection of individual environmental **components**, which is considered 'classical' **technological** protection and protection against adverse **factors**. Potential problems may stem from the fact that integrated approaches, integrated spatial-organisational protection of the environment—the landscape as a whole—are not developed as quickly as desired by professional scientific bodies (Kerényi 2003; Hynek 2009, 2011a).

In spite of the above stated, the spatial approach shows also considerable results. The following facts bolstered development of a spatial conception of the environmental care and facilitated solution of spatial problems:

(a) Development of knowledge: the acceptance that the environment—the living space of humans and other organisms—is composed as a united, complex spatial entity, not just as separate components, therefore, its management needs a united approach. This conception was institutionally recognised at the highest level at the United Nations Conference on Environment and Development in Rio de Janeiro in 1992 (Rio Summit 92), which produced the AGENDA 21. Chapter 10 titled 'Integrated Approach to the Planning and Management of Land Resources' clearly defined such trend as the following:

Land resources are used for a variety of purposes, activities which may compete and lead to conflicts.
But there is only one living space and we all must fit in.

(b) Improvement of 'environmental' technologies and their systematic integration into routine production what caused that today we do not need to fight specially for the cleaning plants, filters, clean technologies, recycling: as they have become a natural part of production. Environmental care has shifted to the platform of sustainable development and it has become a normal, integrated part of technological processes as well as economic, financial and political processes. The active involvement of economy and politics in environmental care dramatically increased.
(c) Regional-spatial policy has become an equal partner to more traditional sectorial development. This also influenced environmental policies and encouraged regionally sustainable development.
(d) Finally, there are still problems which **often cannot be resolved** technologically, they cannot be remedied or bought, e.g. we cannot buy diminishing species, nor ensure extra space for biotops (Hynek 2009) etc.

The key theses of the integrated approach can be formulated, in accordance with Agenda 21 as follows:

- all activities must fit in **the same space** (on the given level: Earth, continent, region, cadastral area);
- if human activities **harm one component of the environment, they will automatically harm the others**. It will also disrupt overall functionality, ecological balance and landscape stability.

The same principles were formulated many times, also in practical works, e.g. as '*all human activities need a place to be realised. While the demand for land increases, supply is fixed. Land, therefore, becomes increasingly scarce. The result is an increase in the number of land conflicts and the level of violence of these conflicts*' (Wehrmann 2012).

But, the same principle also applies that if we institute an optimal measure to protect the landscape as a whole, **all landscape components will be provided protection at the same time**, e.g. if we introduce an optimal distribution of greenery, meadows, pastures, simultaneously we also regulate biodiversity, prevent soil erosion, accumulation of material, water runoff—floods, improve the quality of water, etc.

Appropriate ecological application of the optimal spatial organisation and land use in compliance with the above-mentioned principles will, in **an integrating manner**, ensure:

- satisfactory **ecological quality** of the territory (maintenance of ecological functionality, balance, carrying capacity, stability of the landscape and landscape and biological diversity), in particular, by securing a sufficient area and spatial structure of ecologically stable landscape elements (forests, greenery, grasslands, wetlands, water bodies), i.e. the territorial system of ecological stability;
- protection and rational exploitation **of natural resources** especially soil, water and forest resources, health and recreation resources (protection against erosion, accumulation, pollution), acquired by optimal arrangement of land-use elements;
- protection of immediate **human environment** (maintaining the basic hygiene, health and psychological requirements of the population with regard to the environment), by means of interactive arrangement with anthropic, semi-natural and natural elements (distributions of threatening and threatened elements in space, use of sanitary effects of greenery, availability of recreational areas, aesthetic aspects).

It also needs to be emphasised that the optimisation of spatial organisation and utilisation of landscape **is not** technology in itself and thus **it does not require any investment costs**. It comprises planning and decision-making, which are inherently ***intellectual activities***!

The definition of the ecologisation of the spatial organisation and utilisation of the landscape as one of the main principle of the first Slovak environmental politics (Miklós 1991) was a decisive step for implementing the LANDEP and TSES as obligatory landscape-ecological base documentation to the real spatial planning processes, in particular, to the territorial planning and agricultural land arrangement projects. After this, logically followed that spatial planning processes were considered as the tools for the environmental care with regard to sustainable development, as an organic part of the complex development politics of the state. So, the spatial approach became the frame principle for all the nature conservation, the improvement of the ecological stability of the landscape, the optimum organisation and utilisation of the agricultural landscapes, the integration of the management of natural resources, in particular integrate management of waters, forests and soils with nature conservation, the integration of the landscape-ecological issues to complex spatial/territorial/physical plannings.

The real tasks and real tools of the spatial approach to the environmental care are:

1. **conservate** the valuable landscape structure
   ➤ **nature conservation**
2. **improve the ecological stability** of the landscape
   ➤ **ecological networks planning**
3. **promote the optimum organisation** and utilisation of the agricultural landscapes
   ➤ **land-use planning, land arrangement**
4. **facilitatethe integration** of the management of natural resources
   ➤ **integrated landscape management, including the management of natural resources**
5. **integrate** the landscape-ecological issues to complex spatial planning
   ➤ **landscape ecological plans** in spatial /territorial/physical plannings

We still consider the well developed, legally supported and methods of LANDEP and TSES implemented to practice a base for the ecologically optimal spatial organisation and utilisation of landscape, as such for the base for the sustainable integrated landscape management, and, consequently a base for the spatial/complex protection of the environment (in contrast to the technological protection of the individual components of the environment).

We may state that:

- Ecologically optimal spatial organisation and use of the landscape constitute a principle of integrated environmental protection.
- Landscape-ecological planning is a tool for spatial/integrated environmental care and protection.

Table 1.1 Influence of ecologically optimal spatial organisation of the landscape

	Main spheres of influence			
	Optimum organisation of natural gravitational regions	Protection of landscape-ecological stability	Protection of natural resources (especially soil, water, forests)	Protection of immediate human environment
Tools and methods	Exploiting the gravitational energy of position for production activities	Developing the territorial system of ecological stability, localisation of biocentres and biocorridors	Optimum spatial organisation of landscape elements in river basins	Optimal localisation of residential and recreation areas with respect to threatening sources
	Harmonising the inherent gravity with socio-economic movement	Preservation of historical landscape structures	Optimum localisation of stabilising elements for various types of slope dynamics	Elimination of the neighbouring threatening phenomena
	Localisation of activities by energy and material balance of natural regions	Localisation of specifically protected areas and their linkage to the network of protected areas	Localisation of natural stands, optimisation of forest management processes	Arrangement of protection, sanitary, aesthetic greenery with regard to settlements
Impact	Reducing energy inputs for transport and production Transportation time savings Improving accessibility of services	Protection of genetic resources Biodiversity conservation Preservation of ecological stability Maintaining the conditions for biological pest control Reducing the amount of chemicals	Protection against soil erosion and accumulation in rivers and reservoirs Surface water quality protection Stable water balance Protection of forest production capacity	Protection of physical health Beneficial effect on human mind: environmental well-being Aesthetic impacts

So, the integrated landscape management performs the task of protection and harmonisation of human activities with the conditions of the landscape as a geosystem. In general, the relations of the tools of spatial organisation of the landscape with their impact and sphere influence are schematised in Table 1.1.

1.3 Institutional Instruments for Integrated Landscape Management in Slovakia

The instruments for spatial arrangement of landscape are represented by planning processes. Actually, each spatial planning process bears common basic aspects and goals, which basically can be marked as the optimum spatial organisation of a territory.

The following paragraphs will provide a description of selected planning instruments in Slovakia and their relation to the ecologically optimal organisation and utilisation of landscape.

Several of the integrating-integrated instruments for spatial management are already included to Slovak legislation.

The instruments falling under the remit of the Ministry of Environment of Slovak Republic are:

- ***nature conservation and landscape protection***;
- ***watershed and river basin management***;
- ***flood protection***;
- ***ecological aspects of spatial planning***.

Another integrating tool within the competence of this ministry is the integrated pollution prevention and control (IPPC), and environmental impact assessment (E.I.A), which includes control mechanisms of the process of spatial planning.

The Ministry of Transport, Construction and Regional Development administer the most comprehensive tool promoting the integrated approach to spatial organisation, namely:

- ***territorial (physical) planning***.

Territorial planning represents the most capable integrating spatial tool.

The competencies of the Ministry of Agriculture and Rural Development relate to:

- ***agricultural land consolidations projects;***
- ***forest management plans***.

A significant integrating element of all the above-mentioned planning procedures is the **territorial system of ecological stability** TSES acting as a mandatory part in all these procedures (see below).

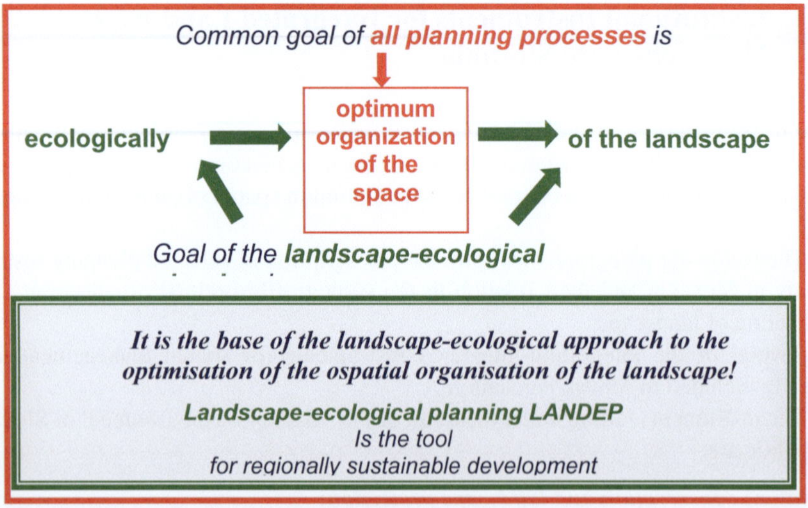

The landscape-ecological and integration aspects of the institutional instruments listed above can be characterised as follows (Miklós et al. 2011b).

1.3.1 Nature Conservation and Landscape Protection

Current integrated nature protection is promoted especially by institutionalised instruments of territorial/spatial nature conservation, often referred to as small-scale or large-scale nature conservation of **conservation areas**. The main supporting documents are:

- Conceptions of landscape and nature protection of different periods.
- Act on **Nature Conservation and Landscape Protection** No. 543/2002 Coll. as amended later, and its implementing regulations. The sections discussing spatial planning processes also define:
  - the territorial system of ecological stability TSES (biocentres, biocorridors, interactive elements, ecostabilising measures);
  - the national network of protected areas;
  - the European network of protected areas Natura 2000 (Areas of European significance, Protected bird areas).

  The nature conservation is defined, of course, also in provisions of other acts (e.g. Water Act, Forest Act, Act on monuments).

Conservation of areas has inherent character of spatially motivated integrated management because it relies on the legal restrictions concerning activities of **all sectors** in the landscape. The problem is that the integrating character is often perceived only as an aspect, which impedes economic development.

The integrating aspect of nature conservation is most evident in the concept of TSES, in particular, in the part discussing ecostabilising measures. It should be said that traditional nature protection relies on the traditional legal protection and nature-conservation methods despite the fact that they bring many conflicts concerning the users of protected parts of the landscape. The major challenge is—and the most important task with regard to integration—to effectively implement the TSES into other spatial planning mechanisms.

Modern nature conservation should also apply the concept of representative geoecosystems to ensure a continuous assessment of the subject, boundaries and level of protection, and thus enhance the management of protected areas (Miklós et al. 2006).

1.3.2 River Basin Management

The landscape-ecological principle of integrated river basin management is relatively simple and can be characterised as follows:

(a) Water is a formless fluid. Its form of appearance, quantity, quality, location as well as surface and underground movement is determined by the gravitation, surface quality, geological layers and climate conditions, therefore

(b) water management should not only begin in a riverbed or groundwater deposits, but we need to look after the '**vessel**' where water accumulates and flows. From the geosystem point of view, the vessel is the whole **landscape as a geosystem**. In other words,

   - where there is a river basin, there is landscape,
   - the landscape does not exist outside river basins.

(c) Water management is traditionally bound with watershed divide—is understood as the limits of spatial landscape subsystems, which delineate the direction of water runoff on the surface.

These principles represent a significant shift from technical water-management view to integrated conception of water management, although water managers have certain reservation in this respect.

The legal basis of integrated river basin management is based on the **Water Framework Directive**—WFD (European Parliament and Council Directive 2000/60/EC) establishing a framework for community action in the field of water management). WFD is in Slovakia reflected in the Act No. 364/2004 on Waters, as amended later (Water Act). WFD and the Water Act stipulate:

- coordination of planning throughout the river basin,
- comprehensive management of sources of surface and groundwater.

They explicitly require the coordination of water management tasks with the development and utilisation of the **water protecting function of TSES**.

Specific instruments to secure the principles include:

- Strategy for the Implementation of Water Framework Directive in the Slovak Republic
- Water Plan of the Slovak Republic (Vodný plán Slovenskej republiky 2009),
- Management Plan of the Basins of rivers Danube and Wisla as actualised in 2015 (Plán … 2015) and management plans of partial river basins (Plány … 2015).

1.3.3 Territorial/Spatial Planning

Territorial/spatial planning represents the most integrating concept with respect to the aforementioned principles as well as the principles set out in AGENDA 21, therefore, we will discuss it in detail.

To specify the term 'territorial' planning: it was born as the planning of urbanisation and construction activities. Later on, it has broadened its scope according to the Act No. 50/1976 Coll. on Territorial Planning and Building Code. The act defined the scope of territorial planning

- for **the entire territory** (there are no white patches);
- for optimal division of the territory to accommodate **all activities**.

In 1989, the act became the part of environmental legislation under the auspice of the Ministry of Environment of Slovakia, in particular, by amendments No. 103/1990 and No. 262/1992. Even more noticeable 'greening' of the law was established by a major amendment to the Act in 2000 (Act No. 237/2000 Coll.). The amendment defined the objective and task of the territorial planning as 'spatial arrangement and functional utilisation of the territory'. So, by its character the territorial planning is partly the planning of the urbanised territories up to master plans of communities and their functional zones, partly deals with the non-urbanised territory, in this part, has the character of land-use plan.

(§ 1) of the Act No. 50/1976 Coll. as amended by Act No. 237/2000 Coll.
(1) Territorial planning systematically and comprehensively addresses the spatial arrangement and functional use of territory, lays down its principles, it proposes the material and chronological coordination of activities which influences the environment, ecological stability, cultural-historical values of the territory, landscape development in accordance with the principles of sustainable development.

After 2002, the territorial planning has been divided. Pursuant to act on competences the '*territorial planning except its ecological aspects*' fell under the remit of the Ministry of Construction and Regional Development, from 2010 it was shifted to the competence of the Ministry of Interior, later came under the Ministry of Transport, Construction and Regional Development. The '*ecological aspects of territorial planning*' remained within the competence of the Ministry of Environment. The ecological aspects of territorial planning include mainly landscape-ecological planning (LANDEP) and the territorial systems of ecological stability (TSES).

After 2006, territorial planning—as an instrument of environmental care—gradually started losing its importance, especially at the level of central authorities. The competencies of territorial planning were, at the regional level, passed on municipalities which act as building offices. However, municipalities exhibit rather reserved stance to territorial planning especially due to its cost as well as complications related to their approval process. Therefore, municipalities resolve territorial problems mostly by land-use decisions, which does not always take account of all ecological aspects.

Anyway, the territorial planning performs as a true integrating tool, as prescribed by AGENDA 21.

> ***Agenda 21, Chapter 10***:
> Objective and outcome of the planning and management of natural resources:
> Integrated physical planning and management
> must be the **framework and basis**
> for each type of sectoral planning

1.3.3.1 The Landscape-ecological Aspects in Territorial Planning

The most important ecological-optimisation aspects in territorial planning Act (Act No. 50/1976 Coll., as amended by Act No. 237/2000 Coll.) can be summarised as follows:

- critical definitions are based on landscape-ecological theories and methodologies, such as:
  - landscape as a geosystem;
  - properties of the elements of landscape structure as regulative which serve as limits, restrictions or supporting factors for utilisation of landscape by man;
  - ecologically optimal spatial arrangement and functional use of landscape (landscape-ecological plan) as a complex process of mutual harmonising of spatial requirements of economic and other human activities with the landscape-ecological conditions,
- landscape-ecological plan (LANDEP) is a functional part of the step named 'Surveys and analyses'
- ecologically optimal utilisation of landscape is defined as a binding regulation of utilisation and spatial arrangement of territory;

- territorial system of ecological stability TSES or its elements (biocentres, biocorridors, interactive elements, ecostabilising measures) are also defined as an obligatory regulation.

The following paragraphs outline excerpts of the Act No. 50/1976 Coll., as amended by Act No. 237/2000 Coll., which marked the most significant still valid 'greening' of the Act. We focused on the legal provisions of the following areas:

- Binding part and guiding part of land-use planning documentation,
- Landscape ecology in land-use planning,
- Landscape in the act,
- Properties of landscape elements in the act—regulations,
- Landscape-ecological plan in the act,
- Ecologically optimal arrangement of the territory and landscape elements in the act.

Wording of the selected articles is given in italics.

Binding part and guiding part of territorial planning documentation

§ 13 (1) *The territorial planning documentation consists of* ***binding*** *part and* ***guiding*** *part. In the Overall plan for development of Slovakia, there are established as binding the* ***principles and regulatives*** *that guide the requirements especially of sectoral plans …with accordance to* ***sustainable development, environmental protection, natural and cultural heritage****.*

(2) *The approving authority determines the binding part and guiding part of territorial planning documentation.* ***In the binding part****, it always determines public buildings and* ***protected segments of landscape****.*

Landscape ecology in territorial planning

§ 1 Objectives and tasks of territorial planning

It is informative to compare the article valid until 31.7.2000 which read:

(1) *territorial planning continuously and comprehensively addresses the functional utilisation of territory, lays down the principles of its organisation and the material and chronological coordination and other activities influencing development of the territory.*

(2) *territorial planning creates the conditions for permanent harmony of all natural, civilisation and cultural values in the territory, especially with respect to the care for* ***the environment and protection of its main elements—soil, water and air****.*

since 31.7. 2000 they read:

(1) *territorial planning systematically and comprehensively addresses the spatial arrangement and functional use of territory, lays down its principles, it proposes the material and chronological coordination of activities which influence*

environment, ***ecological stability***, *cultural-historical values of landscape, landscape development in accordance with the principles of* ***sustainable development***.

(2) *territorial planning creates the conditions for permanent harmony of all activities in the territory, with particular regard to care for the environment, reaching the ecological balance and ensuring* ***sustainable development, considerate utilisation of natural resources and protection of natural, civilisation and cultural values***.

Other significant provisions currently in force:
§ 2 (1) *territorial planning includes these tasks and activities:*

(a) *it determines* ***the regulations of spatial arrangement*** *and functional land-use,*

…

(c) *it defines* ***protected areas, protected buildings, quiet and protection zones (hereinafter as 'protected areas of landscape')***, *unless they originate under other regulations, and* ***ensures the protection of all protected areas of landscape***,

…

(g) *it determines* ***the principles of the use of natural resources, conditions of the territory and entire environment*** *in order that the activities within it do not exceed the acceptable burden on land (1a) so that there be created and preserved the* ***ecological stability (1b) of the land***,

…

(j) *proposes the land-technical and organisational measures necessary for* ***improvement of environment, achievement of ecological stability and ensuring the sustainable development***.

Landscape in the Act

§ 139 Terms of land-use planning

(3) **Landscape is a complex system of space, location, georelief and other mutually, functionally interconnected material natural elements and elements modified and created by a man**, in particular, the geological substrate and soil creating substratum, water bodies, soil, flora and fauna, artificial objects and the elements of utilisation of territory, as well as their **connections determined by socio-economic phenomena**. ***The landscape represents the environment of man and other living organisms***.

Properties of landscape elements in the Act—regulatives

§ 139 Terms of territorial planning

(1) *Pursuant to the act,* **a regulative** *of spatial arrangement and functional use of landscape is* **a binding directive, which** *guides the location and arrangement of a certain object or implementation of a certain activity in a territory.* It is expressed by the values of **properties of elements of landscape structure** in words, numbers and, if applicable, also graphically. *The regulative has a character of bans, restrictions or supporting factors in relation to* ***spatial arrangement and functional use of the landscape****. Hereby, the regulative determines a* ***forbidden, restricted and allowable activity*** *or function in a territory.*

(2) ***Spatial arrangement and functional use of landscape*** *represents the complex process of mutual harmonising the requirements of economic and other activities of a man in environment.*

Ecologically optimal arrangement of the territory, landscape elements and landscape-ecological plan in the Act

§ 19c Surveys and analyses

(1) *The aim of surveys and analyses is especially, based on territorial planning materials and other binding materials, to acquire the knowledge on condition and possibilities of spatial arrangement and functional use of landscape, determination of problems and mutual interests in the addressed area necessary for producing the commission, draft and proposal of territorial planning documentation.*

(2) ***Within the surveys and analyses it is processed*** *for the territorial plan of region and territorial plan of the municipality the* ***ecologically optimal spatial arrangement and functional use of landscape*** *taking into the account the* ***landscape-ecological****, cultural-historical and socio-economic conditions (hereinafter only '****landscape-ecological plan****').*

§ 139 Terms of territorial planning

(4) ***Ecologically optimal spatial arrangement and functional use of landscape (landscape-ecological plan)*** *is the complex process of mutual harmonising of spatial requirements of economic and other activities of man with* ***the landscape-ecological conditions*** *that result from* ***the landscape structure****.*

… Ecologically optimal spatial arrangement and functional use of landscape at the same time ensures the ***suitable stability of spatial structure of landscape, protection and rational use of nature, biodiversity and natural resources, creation and protection of the territorial system of ecological stability and immediate environment of man****.* ***Landscape structure and its elements*** *are showed as* ***limits, restrictions or supporting factors*** *of required activities in the given territory.*

Table 1.2 Economic and physical categories in planning

Economic categories	***Physical categories***
Region: gravitation spatial unit with elements more strongly bound within this unit thank with elements outside this space	***Territory***: general representation of physical space on the on the earth's surface
Regional policy: transfer of resources from regions with the surplus of resources to regions with the lack of resources	***(spatial policy)***
Regional development: economic and social development of **region as a whole** (vs. **sectoral** development), development and complementarity between different industries and sectors	***Territorial development***: balanced use of physical space in accordance with the **conditions of the primary, secondary and tertiary landscape structure** (abiotic, biotic, and land use)
Regional planning: tool of regional development: planning of the **development of economic and social resources** in the region (transfer, internal development)	***Territorial planning***: tool of environmental care: ecologically optimal planning of the **physical arrangement of** objects and activities on the earth's surface
Demands: development, efforts to overcome limits (victory over nature) = **economic interest**	***Offers***: limits, restrictions, carrying capacity = interests of **environmental protection**

Ecologically optimal spatial and functional arrangement of the territory = landscape-ecological plan

In many countries, territorial planning is given all sorts of names, often physical planning, land-use planning, spatial planning or regional planning. In Slovakia, regional planning has an entirely different character; it is an economic activity. The relation between regional planning and territorial planning is outlined by following terms (Table 1.2):

1.3.3.2 Specification of 'Landscape' Planning and 'Landscape-ecological' Planning

There is no legally defined term 'landscape planning' in Slovakia.

Recently, also the research and assessment of landscape image and character as well as changes in utilisation of land and management of historic landscape structures have attracted a lot of attention and popularity among ecologists, geographers and other related professions, especially with respect to the implementation of the European landscape convention (European … 2000). It focuses on the overall external manifestation of the landscape, its spatial arrangement, its assessment of the 'quality' for living, recreation and various other ways of use resulting from this structure. In principle, such trends draw on the works of Forman and Godron (1981), but the original 'structure—process' approach was later modified, sometimes lacking the landscape-ecological facets, not requiring an application of landscape-ecological approach and qualifications in the field, concentrating on cultural-historical, and also artistic aspects (Marsh 1997; Grodzinski 2005; Ružička et al. 2010; Belčáková 2013a; Rega 2014). Their manifestation can be observed in the development of 'landscaping', landscape architecture. However, these processes are not yet legally defined neither systematically incorporated in the planning instruments of the Slovak Republic. It may even seem that the practitioners of such concepts of 'landscaping' do not desire to be bound by any legislative norms. This could cause that the practical landscaping will be driven by spontaneity, coincidence and the goodwill of participating entities.

As already mentioned the 'landscape-**ecological**' planning is currently included in the legislation of territorial planning. Since the territorial planning is in present divided under the remit of two ministries, it is far not an ideal situation. Therefore, there is a permanent endeavour of landscape ecologists and environmentalists to develop and adopt a new act under on 'landscape-ecological planning' under the auspices of the Ministry of Environment. According to these efforts and previous attempts, the new landscape-ecological plan should represent:

- an integrated ***spatial projection*** of all interests regarding **the protection of nature, landscape and environment**;
- unified ***summary of binding regulations*** issuing from the landscape structure;
- materially defined base for the assumption and acceptation of cultural-historical and aesthetical values in the spatial plannings, and, as such an;
- consolidated unified material for all types of spatial planning processes.

So this landscape-ecological plan should be a complex document which should replace the landscape-ecological plan defined in the currently valid Act on territorial planning but should include also the aspects of character, aesthetics and other values of the landscape.

In principle, it can be said that in the process of environmentally optimal organisation of the landscape:

- landscape-ecological planning LANDEP serves as a basic technical documentation for territorial planning, at the initial stages of the planning process;
- territorial planning based on LANDEP constitutes a skeleton and a fundamental technical instrument of such optimisation,
- landscaping is the final phase of the process, a part of territorial planning documents and spatial planning outputs.

So, the landscaping (landscape architecture, landscape planning) could potentially represent the last stage of ecologically optimal organisation of landscape which could complement a territorial plan from the landscape-aesthetic point of view. But, it is necessary to underpine this process by *landscape-ecological functional principles*, which should promote a more objective application of perception and aesthetic values of the landscape to the plans (Wehrmann 2012).

1.3.4 Agricultural Land Consolidation and Arrangement Projecting

Landscape ecologists see the process of land consolidation as a significant tool to implement ecological principles into the management of agricultural landscape. Also, the primary principles of territorial systems of ecological stability (TSES) were based on the completion and improvement of remnants of greenery along field roads, hedgerows, wetlands in the agricultural landscape, which were named as the 'skeleton' of ecological stability (Lőw et al. 1984; Buček et al. 1986). This skeleton should be integrated with newly projected elements during land arrangement projecting what results to the creation of '**green landscape structure**', '**green infrastructure**', a form of spatial system of linear and small-scale greenery in agricultural landscape to sustain ecologically stable landscape. For that reason, the concept of TSES was incorporated in 1991 into the design process of land consolidation projects (Slovak National Council Act No. 307/1992 on the Protection of Agricultural Land and Act No. 330/1991 on Land consolidation, Land Property Arrangements, Land Offices, Land Fund and Land Associations as later amended and still valid.

Land consolidations fall within the competencies of the Ministry of Agriculture and Rural Development of the Slovak Republic. Land consolidations projects (LCP) are considered an integrating tool for landscape management mainly because the entire agricultural landscape lies within its competence. The territorial system of ecological stability plays an indispensable role in LCP:

- TSES is a mandatory part of LCP;
- increase in ecological stability may be a reason for initiation act of land consolidation project;
- the elements of the TSES and important landscape elements are, in the context of land consolidation, considered common facilities (Act No. 7/2010 Coll. on Flood Protection).

So-called 'ecostabilising measures' are a component of TSES as stipulated by the Act on Nature Conservation (Act No. 543/2002 Coll.) as well as described in the methodological guidelines for TSES projecting (Metodické pokyny … 1993; Izakovičová et al. 2000). Thus, the TSES regulates the spatial arrangement of stabilising elements in the landscape (biocentres, biocorridors and interactive elements) as well as whole-area optimisation of land utilisation. However, the TSES is not sufficiently realised in the practice, even though the concept was proposed, in form of a **green landscape skeleton**, by agricultural planners and landscape ecologists. One serious, albeit pedestrian, problem of LCP is inadequate funding and substantial time and financial costs required to resolve the issues arising from ownership relations and land arrangement, and ultimately there is no time, energy or money left for TSES implementation.

1.3.5 Forest Management Planning

Forest management planning falls under the competence of the Ministry of Agriculture and Rural Development. Landscape ecologists also consider forests a key component of the ecological stability of the landscape because of their biological nature and extensive areas. Therefore, each forest exploitation measure may considerably influence the ecological stability of the entire landscape. Forest management policies declare the importance of forests in each relevant documents, emphasising also their non-production (utility, all-society) functions mostly related to protection of nature and natural resources. According to National Forest Programme (2007) the forest industry made up only 0.59% of GDP in 2005, but together with its utility functions it was estimated as 3.03%. In 2008, the share of GDP was only 0.42%, in 2015 only 0.37%, the figures yearly decrease (Report … 2016). So, the other utility functions of forests many times outweigh the economic importance of timber and other forest goods production. Due also to these facts, we consider ecologically optimal forest management an integrating element in the landscape management.

Specific management actions are governed by forest management plans called Programs of the Care on the Forests (PCF, Act No. 326/2005 Coll. on Forests), which fall under the remit of the Ministry of Agriculture and Rural Development of the Slovak Republic. PCFs regulate differentiated forest management, which must consider various natural, stand, economic and social conditions constituting a basis for declaration of **productive** forests, **protective** forests and **special-purpose** forests. The last two are actually the bearers of mentioned non-production functions of forests. PCFs must also contain measures for nature conservation, which provokes a lot of disputes, because, despite all the ecological declarations in the relevant documents, fundamental awareness as well as knowledge of a forest manager are set to pursue primarily the aspects of production and economic exploitation of wood. Fortunately, timber production is a long-term process and requires a number of measures which must follow also the natural conditions, so, in spite of all those problems and disputes, forest management is an ally in integrated landscape management.

1.3.6 Flood Protection

For landscape management, the Act No. 7/2010 Coll. on Flood Protection is of particular importance. This act has truly an integrating character and its provisions reconcile landscape-ecological information basis, all above-listed legal instruments and explicitly defined the integrated landscape management. In addition, they specify the deployment of flood prevention measures to ensure water retention in the landscape—especially by exploiting the elements of TSES contained in the land consolidations and land-use planning projects. The most important provisions of the act in original wording read:

§ 4 *Measures for flood protection*

(2) *Preventive measures for flood protection include:*

(a) *measures that decelerate runoff from river basins into watercourses, increase retention capacity of river basins and promote natural water storage ... such as regulations in forests, farmland and urban areas,*

§ 5 *Preliminary flood risk assessment*

(2) *... is also an assessment of runoff conditions as well as conditions for increasing the retention capacity of the landscape. As the base document shall be elaborated a set of spatial data named* ***Landscape-Ecological Basis of Integrated Landscape Management*** *on relevant abiotic complexes in river basins, in particular morphometric characteristics of the relief, physical properties of soil and bedrock, as well as spatial data on the elements of the current land-use.*

§ 8 *Flood risk management plan*

(4) *To ensure preventive, flood protection* measures, *according to the spatial data from the Landscape-ecological Basis for Integrated Landscape Management pursuant to § 5 par. 2* ***a Proposal of measures for long-term river basin management*** *shall be developed.*

(5) *the proposal of measures for a long-term river-basin management should also project the* ***TSES, important landscape elements and the eco-stabilising measures...*** *Furthermore, the proposal shall include spatial projection of nature protection areas, protected areas of water resources, land resources, forest land resources and mineral resources. Detailed content and scope of the Landscape-ecological Basis for Integrated Landscape Management and Proposal of measures for long-term river basin management will be specified by the Ministry by means of generally binding regulations.*

(6) *Proposals of measures for long-term river basin management are a part of the* ***Flood Risk Management Plan*** *and the basis for land consolidation projects, land-use plans and land-use decisions.*

The most interesting provision of the act was § 9 on the coordination of different planning tools what red: *plan of the flood risk management and the river basin management plan shall be coordinated with the land arrangement projects, territorial plans and forest management plans. They alltogether will constitute the tool of* ***integrated landscape management*** *on the whole territory of the river basin.*

Although the original wording of the provisions might be changed by later amendment, the spirit of integrated approach remained.

1.3.7 Integrated Spatial Information Systems

One of the crucial conditions for further development of integrated approaches to the landscape management is building an integrated spatial information system applying **GIS** technologies. So far, no such system existed, which was caused mostly by various technological and organisational problems, but also subjective aversion of different organisations to design a unified and for all sectors accessible information base. Common EU legislation might resolve this situation, in particular, by the INSPIRE Directive 2007/2/EC (INSPIRE—INfrastructure for SPatial InfoRmation in Europe), which obliges the EU Member States to run a spatial information system on a unified cartographic basis, with unified information (e.g. list of objects) based on GIS technologies. The directive was implied to the Act No. 3/2010 Coll on National Infrastructure for Spatial Information (NISI).

Since 2000, the Ministry of Environment has been committed to developing such a system, as well as feeding it with required data as a part of the Landscape-Ecological Basis of Integrated Landscape Management LEB ILM. After its completion, the basis will allow, based on uniformly processed file, to manage and make decisions regarding practically all partial outputs of the integrated landscape management.

To satisfy the needs of these procedures, which fall under the Ministry of Environment of the Slovak Republic, the objectives of LEB ILM were formulated as follows:

- to create **multifunctional basis documents** for all program objectives of the MoE of the Slovak Republic (analytical database, abiocomplexes, landscape-ecological complexes)
- to apply the materials within **integrated landscape management** to:
  - identification of representative geoecosystems,
  - concept of protection and integrated management of protected areas,
  - identification cards of conservation areas, border revision, level of protection,
  - ecological aspects of territorial planning—landscape-ecological planning, TSES,
  - integrated river basin management (river basin management plans),
  - management of the flood risk.

The most important provisions of the Act No. 3/2010 Coll on National Infrastructure for Spatial Information in terms of integrated landscape management are as follows:

§ 1 Subject of the Law

This Article provides, inter alia,

(a) *establishment of a national infrastructure for spatial information for the purposes of management and decision-making in public administration,*

(b) *coordination of development and operation of the national infrastructure for spatial information,*

(f) *establishment of a national geo-portal.*

§ 8 National Geoportal

This section addresses the accessibility to NISI services as follows:

(1) *The obliged person … is obliged to provide access to network services, as well as spatial data through the National Geoportal. The obliged person may provide access to these services as well as spatial data also through their own access points.*

(2) *The National Geoportal can allow access to spatial data and spatial data services to third parties, if they satisfy the conditions set out ….*

The content of NISI is defined by the INSPIRE Directive and detailed content should be specified by a decree of the Ministry of Environment of Slovak Republic.

1.3.8 Interrelation of Institutional Instruments of Integrated Landscape Management

Legally supported instruments of integrated landscape management can be divided, according to principles defined by AGENDA 21 and later documents, into four main groups as:

Information base of integrated landscape management

Information base is defined by Act No. 3/2010 Coll. on NISI and Act No. 7/2010 Coll., which also defines **LEB ILM**. Cartographic base consists of basic geodetic and cartographic documents—maps and the document called basis for geographic information systems (Kolektív autorov 2008; Koncepcia … 2006). The system is complemented by subject data on LEB ILM, data obtained from GEOFOND, but also various atlas works, catalogues and other spatial data from different sectors. The instruments should form a unified information base for all sectoral planning procedures.

Landscape-ecological framework for integrated landscape management

According to Agenda 21, the skeleton and framework for all sectoral planning procedures should be a physical (territorial, spatial) plan created on a landscape-ecological basis. In view of this requirement, we developed the methodology of landscape-ecological planning, LANDEP, as well as the method of the territorial system of ecological stability, TSES, which together constitute the legally supported base for the process of territorial planning. Thus, the LANDEP, TSES and territorial planning can be considered an interconnecting component which professionally processes landscape-ecological data for territorial planning and other sectorial planning procedures.

Sectoral spatial planning instruments

These instruments ensure final implementation of landscape-ecological aspects into realisation practice. They include: projecting of land consolidations, forest management planning and projecting, integrated management of conservation areas, water management planning, river basin management, flood risk management and other sectoral plannings—e.g. exploitation of minerals, recreation and tourism. Coordination of these instruments following in appropriate sequence represents **the integrated landscape management**.

Instruments for control and assessment of the impact of planning procedures on the environment

These tools are oriented not to a direct management of the landscape, but on the control and assessment of the impact of the spatial activities of the sectors, therefore we can consider them as the important tools for regulation of those activities. Two legal instruments can be mentioned:

- the Integrated Prevention and Pollution Control (IPPC) executed according to the Act No. 39/2013 Coll.
- the Environmental Impact Assesment (E.I.A.) and Strategic Environmental Assessment (S.E.A.) executed according to the Act No. 24/2006 Coll. As later amended.

A specific legally defined integrating element TSES

A specific integrating interrelation element for all those tools is the territorial system of ecological stability TSES, since it is defined and plays important role in each of above-mentioned institutional tools of integrated management of the landscape.

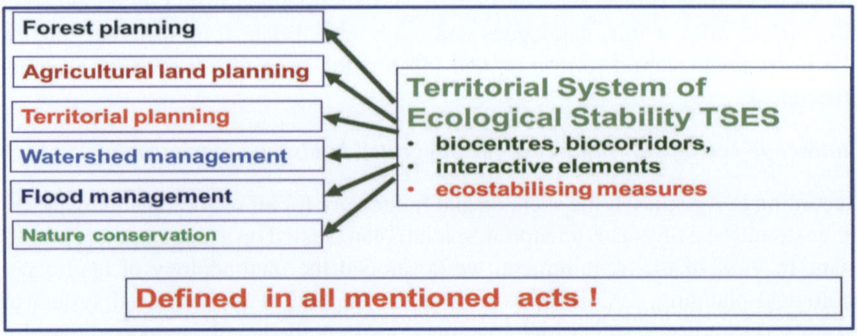

As we explained, the integrated management of the landscape is not a single tool, or a single procedure. The core and the essence of it is the harmonisation of the demands of all sectors towards the landscape (towards the space) by the **harmonisation of all spatial planning, projecting and other management tools**.

This is an extremely difficult task which needs strong factual and timely coordination of all concerned sectoral institutions.

The interrelation and interconnection of those tools show the scheme in Fig. 1.1.

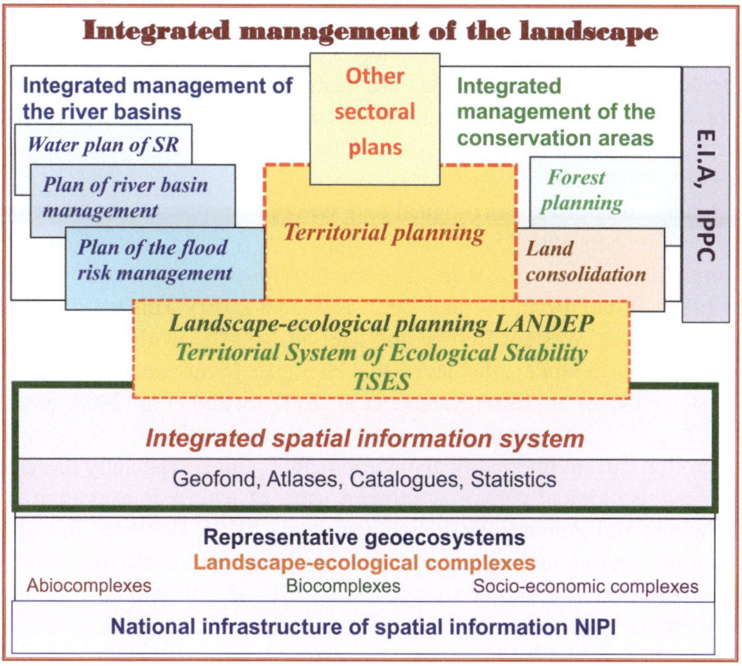

Fig. 1.1 The interrelations of the institutional instruments of integrated landscape management

1.4 The Object of LANDEP: The Landscape as a Geosystem

There were no questions about the basic object of landscape-ecological planning: it is the landscape, of course. Nevertheless, since there are several directions of the of the understanding of the scientific concept 'landscape', the basic aspect of the way how to define the object of the LANDEP was which concept of the landscape supports the functional evaluation of suitability of the landscape for all sectorial activities and for the whole territory of interest, as described in AGENDA 21 and as defined in Acts (see Sect. 1.3)

Without going to deep theoretical analyses of the scientific terms, we can generalise the understandings of the landscape into few main streams:

The approaches defining landscape as a natural complex were and still are popular in Russian/Soviet and East-European landscape-ecological school (e.g. Armand 1952, 1964; Sochava 1977; Isachenko 1981; Preobrazhensky 1983; Grodzinski 2005; Khoroshev 2010; Bastian et al. 2015).

The landscape as a natural–socio-economic complex was characteristic for German and partially, Western European, landscape-ecological school (Neef 1967; Neef et al. 1973; Haber 1980; Snacken and Antrop 1983; Leser 1997; Brandt 1999). This approach has been widely applied in the Central European landscape ecology school

(Verrasztó 1979, Drdoš et al. 1980; Mičian 1982; Csorba 1988; Richling and Solon 1993, Lowiczki and Mizgajski 2013; Veteikis et al. 2015).

Both above-mentioned approaches are based on in-depth vertical and horizontal comprehensive analysis and synthesis of natural and socio-economic section of the geosphere, and both are predecessors of the geosystem approach (Krcho 1968; Demek 1974, 1978)—see below.

The American landscape-ecological and Western European schools relied, and still rely, on the characteristics of the landscape with regard to the structure of land cover, originally with emphasis on the structure-function approach (Forman and Godron 1981; Turner 1990; Forman 1995; Nassauer 2012). This approach became very popular also in Western Europe and later all over the world. Huge amount of landscape-ecological studies displayed this character (Schreiber 1980; Brandt and Vejre 2004; Wrbka et al. 2004; Kienast et al. 2007; Szabó et al. 2008; Štefunková and Hanušin 2015). This popularity contributed to the fact that the landscape issues entered into the current international environmental policy, especially its ecological segment—e.g. ecological networks, modern ways of nature conservation (Langer 1990; Jongman and Pungetti 2004; Schilleci et al. 2017). However, this approach raises a lot of methodological uncertainties and scientific misinterpretations when a number of studies concentrate merely on the structure of land cover only, 'even when applied in a more amateurish way' (Antrop 2013). This approach is abundantly applied in landscape architecture (landscaping).

The holistic definitions of landscape resulted into more or less philosophically tuned theories, which often lack a precise definition of the object, and even reduced the landscape to an aspect, reflection of the real world (e.g. Zonneveld 1981; Golley and Bellot 1991; Hynek 2010; Hunziker et al. 2007). This group includes the perception of the entire human ecosystem (Naveh and Lieberman 1993), 'three-pronged' view of the landscape as 'scenery—pattern—system (Zonneveld 1995), or even the five-dimensional' spatial–mental–temporal–natural–cultural complex 'system perception of the landscape' (Tress and Tress 2001, 2002). Nevertheless, whatever holistic is the landscape, for practical purposes needs some concretisation (Verrasztó 2017; Antrop and van Eetvelde 2017). It is to mention that in reality whatever holistic approach finishes at the end of the day to a much more simple process than a holistic one. The problem is when even the simplified physical–geographical complexity is abandoned. The results, then, represent only analyses of some relation. Such approach cannot be considered as either comprehensive or holistic.

Of course, the differences in the understanding of particular studies and projects are not as clear-cut and the distribution of these approaches is not strictly demarcated (Hynek 2011b, Wu 2013). The above-mentioned classification should serve as an introduction for those, who want to further study the theoretical–methodological issues of the landscape. Nowadays, all these approaches might be divided—neither sharp—in two directions: geosystem based, materially conceived landscape structured by all components of geographical sphere (may say 'hard' concepts), and on the land-cover concentrating approaches, often evaluating the perception-based values as the beauty, the heritage and other values ('soft' concepts). This division is never a ranking, just a differentiation which "predestine", "designate" the possibility of application of different concepts in practical planning procedures. Without any doubt a landscape type as a geosystem defined exactly, e.g. by relief dissection, soil depth or a biocoenoses, has a different normative implication for the planning and management of the territory than the characteristics of the aesthetic or cultural values of a landscape, which is much more subjective and changing, e.g. according to the persons, angle of view, etc. Of course, the above stated does not mean at all, that the beauty and other similar values of the landscape are not important indices. The opposite is true: right because of their fuzzy character their implementation needs more diligence, in order not to lose them in design and planning processes.

The differences between two approaches are obvious if one compares the definitions based on general system theory (Bertalanffy 1968), modified for geographical and landscape sciences by many representatives of the first group, such as Neef and others (Neef 1967; Neef et al. 1973; Chorley and Kennedy 1971; Preobrazhensky and Minc 1973; Sochava 1977; Krcho 1968; 1978), with the definition given e.g. by the European Landscape Convention (2000). The European Landscape Convention is an attempt to bring a holistic understanding of the landscape into political sphere. Like each international convention—also this one manifests compromises between professionals, diplomats and politicians.

The definition presented by *European Landscape Convention* reads: 'Landscape is an area, ***as perceived by people***, whose character is the result of the action and interaction of natural or human factors' (Article 1). The wording of the definition itself is not incorrect, however, does not help to introduce the material essence of geosystem into planning procedures. Landscape is defined as an imaginary entity based on **perception** of its character. Other articles of the Convention define the landscape as an assembly of 'heritage', 'values', 'quality' (Article 5, 6). This definition allows very broad interpretations—'as perceived by people'—which could exaggerate inconsistency of the scientific and technical basis for real plans and projects. Another problem of the Convention is that the governments should apply hard management procedures to the softly defined object.

In this concern, it is to underline that each status of the landscape, whatever holistically perceived, the status of its structure, its quality, values, the 'shape/scape', is the result of the land use of each single material spot of the landscape (Zonneveld 1981; Haber 1990, 2008; Khoroshev and Koshcheeva 2009). Those single spots create the 'shape' of the land also in a choric dimension. We can evaluate this shape as more or less attractive, valuable, with lower or higher quality. If we wish to protect or change this shape, the values, the quality, there is no way to do it 'holistically', we have to protect or change **the use** of each respective spot. Furthermore, it is to mention, that each part of the landscape has its owner or user, who has his own imagination on the further shape of this ownership. He can be forced to keep or change the use of his ownership only according to the legal, well-defined normative tools. As stated in the practical guide for land-use planning by Wehrman (2012), the land-use and land ownership are two independent factors highly relevant for development. This circumstance also presses upon the materialistic definition of each spot of the landscape.

Nevertheless, the Convention serves as a very important tool for waking the general awareness of the politics and publics on the landscape issues. The Convention is also the main pillar for the landscapers who consider the landscape as the 'shape/scape' of the land, the cultural-heritage value, or as an aspect, a phenomenon of the real world. The specialists from this group do not always insist on the knowledge of the elements of landscape, of their physical structure (see, e.g. Breuste et al. 2009). Landscape architects, designers prefer many times fuzzy information on landscapes, more than exactly defined indicators of regulative character, because they offer them more freedom, i.e. less obligations. The problem is that the quality of implementation of landscape information is hardly controllable, less normative, as in case of numeric or other exact regulatives.

The above-mentioned never means that this approach has no implementation to practical landscape protection and design. There is a huge amount of good practices with application of landscape values to the design—in the case of proper knowledge of the authors on the landscape (Nassauer 1997; Nassauer and Opdam 2008; Breuste et al. 2009; Mizgajski and Markuszewska 2010; Nassauer 2012; Foo et al. 2015).

In reality, the general public, decision-makers, planners and designers understand under the landscape mostly visual, aesthetic, conservational or cultural aspects of the landscape. Nevertheless, in our countries, management of the landscape and its changes are regulated by legal procedures—by **spatial planning**. Our exercises showed that for the decision on the optimal use of the landscape, according to the real materialistic conditions, we need to define the landscape as a

system of material elements with precisely defined indicators,

which plays the role of a physical **regulative** for the decision-making process (Khoroshev and Koshcheeva 2009; Bastian et al. 2015). If the elements of the landscape are not defined materially, if they are not related to a regulative, then the politicians, the planners, the designers may apply the information on the landscape in an arbitrary way, not as an obligatory regulative dependent on physical properties.

We consider a pragmatic compromise between holistic theories and real research with regard to utilisation of acquired results with maximum complexity is the understanding of the landscape as a geosystem.

The material basis and object of the landscape-ecological planning LANDEP is the

landscape as a geosystem,

a system of material elements with precisely defined indicators.

The general theoretical and methodological aspects of the landscape as a geosystem with regard to practical use of this concept in planning processes are sufficiently presented in the works Miklós and Izakovičová (1997), Miklós and Špinerová (2011), Miklós et al. (2015).

Following chapters outlines only the most essential aspects, which are closely associated with the LANDEP methodology and its particular parts.

1.4.1 The System Theory and the Geosystem

The geosystem theory is based on the basic system theory, and the crucial aspect of this theory is an assertion that **a system is more than a mere sum of its elements**, and on the basic definition that the **system is a set of elements and relations** (General System Theory—GST, Bertalanffy 1950, 1968).

The geosystem theory was developed by gradual adaptation of the German term 'Gestalt' in geographical theory. Many authors tried to express the holistic nature of the landscape. A fundamental material definition for further application of the theory into practice was proposed by Neef: 'landscape is a concrete part of the earth's surface with cohesive structure and similar character of relations *(durch einheitliche Structur und gleiches Character Wirkungsgefüge geprägten konkreten Teil der Erdoberfläche*' (Neef 1967). Many other geosystem definitions emerged in that period (Krcho 1968; 1974; 1978; Chorley and Kennedy 1971; Sochava 1977; Neef et al. 1973, 1973; Preobrazhensky and Minc 1973; Demek 1974; Mičian 1982; Preobrazhensky 1983; Snacken and Antrop 1983; Richling and Solon 1993; Naveh and Lieberman 1993 and many others). Based on these studies, after the analysis of crucial aspects of definitions of a geosystem by various authors a simple, but comprehensive definition of geosystem can be proposed as: 'geosystem is a complex of elements (components) of geographical sphere and their interactions with each other (Krcho 1978; Miklós and Izakovičová 1997; Miklós and Špinerová 2011).

There are, of course, also other definitions of a geosystem, which similarly operate system terms, e.g. structure, pattern of functioning, dynamics, matter, energy, information, synergy, spatiality and temporality. With respect to a simple definition of geosystem, these terms are rendered redundant because they are implicitly encompassed by the terms **system, complex, element and relation**.

1.4.2 Methodical Base of the LANDEP: Application of Models of Geosystems

Models can be regarded as an abstraction of reality. In this respect, the simplest model of geosystems is their perception as a material cut of the geographical sphere (Fig. 1.2).

For the purposes of landscape-ecological planning, we model the landscape as geosystem in two types of models as:

Base for the analyses in LANDEP: the topical model of geosystem

The topical model of geosystem comprises the elements of the system represented by the components of the geographical sphere $\boldsymbol{a}_1$–$\boldsymbol{a}_n$. Relationships within geosystems are labelled $\boldsymbol{r}_n$. The model of geosystem can be formulated as follows:

$$S_{GK} = \{\boldsymbol{a}_n, \boldsymbol{r}_n, \}$$

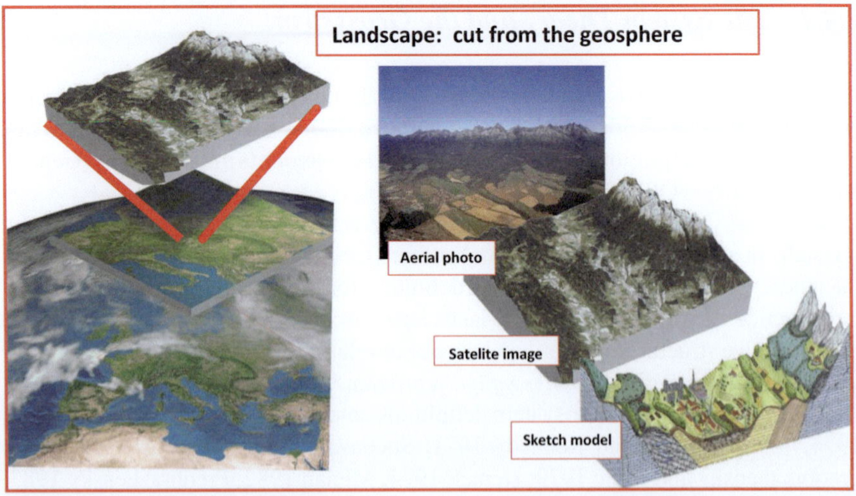

Fig. 1.2 Landscape as a material section of the geographical sphere

In case of applied landscape-ecological studies, a formal transcription of all relations between all elements would be extremely difficult, as we will never identify all relationships. Therefore, for practical reasons, we use geocomplexes, which have similar material essence to geosystems, but they are formally transcribed only as a complex of element/components of geographical sphere as

$$\boldsymbol{G_K} = (\boldsymbol{a_n}),$$

by implicit attribution of a set of relations to a particular complex (Krcho 1968, 1978).

Such topical model is also called a mono-systemic model (Preobrazhensky and Minc 1973; Preobrazhensky 1983).

Topical model of a geosystem (Fig. 1.3) is used as the **principle for the landscape-ecological analyses of the LANDEP** (more details see in Chap. 3).

Base for syntheses in LANDEP: choric model of a geosystem

The elements of the system according to choric model are partial spatial systems/spatial subsystems (Krcho 1974, 1978). The choric model is also called a polysystemic model (Preobrazhensky and Minc 1973; Preobrazhensky 1983) because except for a description of the system $\boldsymbol{SG_K}$ as a whole, each spatial subsystem $\boldsymbol{SG_{(n)}}$ can be also described by means of a topical model $\boldsymbol{G_K}$ introduced in the previous chapter.

Thus, the model can be transcribed as follows:

$$\boldsymbol{SG_K} = \{\boldsymbol{SG}_{(1)}, \boldsymbol{SG}_{(2)}, \ldots \boldsymbol{SG}_{(n)}\},$$

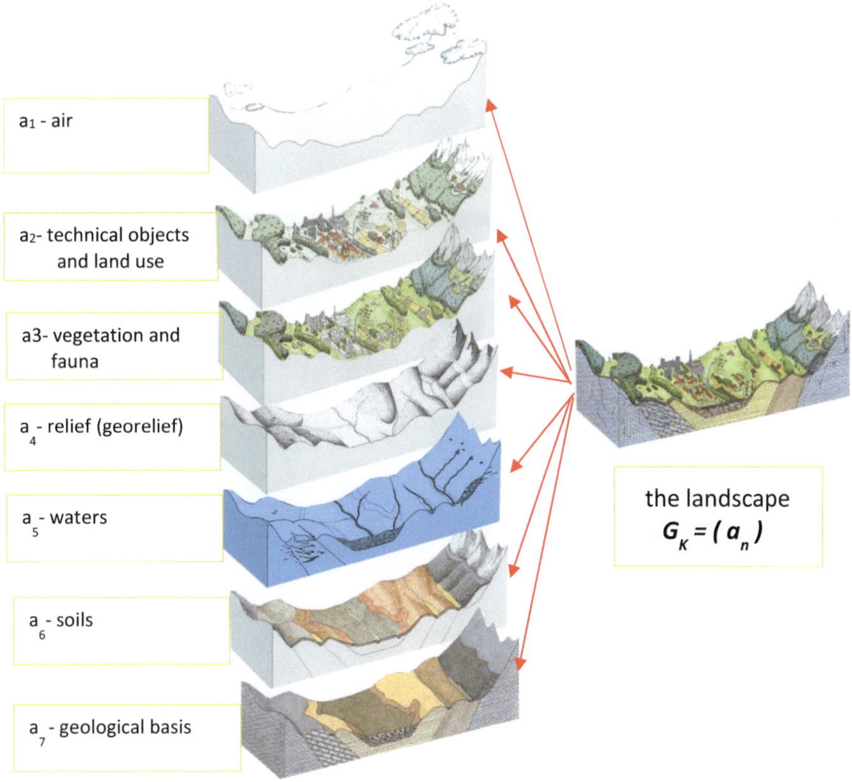

Fig. 1.3 Topical model of geosystem

where the elements of the system are partial spatial systems (spatial subsystems) $SG_{(1)}$ to $SG_{(n)}$. More accurate representation in polysystemic form is

$$S\{G_K(a_n)\} = [S\{G_K(a_n)\}_{(1)}, S\{G_K(a_n)\}_{(2)}, \ldots S\{G_K(a_n)\}_{(n)}].$$

The choric model of a geosystem (Fig. 1.4) is also widely used and constitutes a principle for landscape-ecological **syntheses** in the LANDEP methodology, e.g. creating and characterisation of **abiotic complexes and landscape-ecological complexes** (more details see in Chap. 4).

Base for interpretations, evaluations and propositions in LANDEP: socio-economic factors/phenomena in the landscape

Socio-economic phenomena/factors in the landscape (SEF) are specific **non-material elements** of the landscape structure. They mainly have the character of legislatively designated **zones, protected areas and territories**, which are governed by a range of **restrictions and bans**.

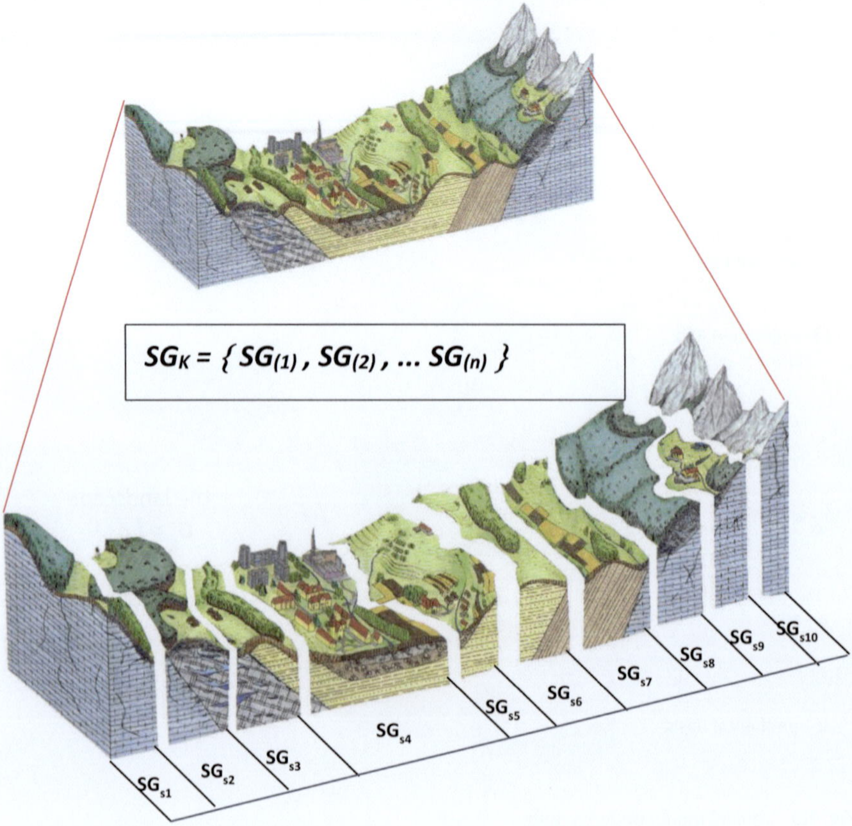

Fig. 1.4 Choric model of geosystem

As they are non-material, intangible, they often overlap each other and according to their character, they are mutually supporting, neutral, harming or even eliminating each other. (Fig. 1.5)

The following groups of the socio-economic factors are present in the landscape:

(a) SEF of nature conservation (b) SEF of protection of natural resources (c) SEF of urban, residential and recreation areas (d) SEF of communication and transport (e) SEF of industrial and mining activities (g) SEF of territorial units—administrative boundaries (h) SEF of deterioration of natural resources and the environment.

In spite of the SEF being non-material, they have decisive influence on the spatial planning processes since they are institutionally expressed **interests, manifestations and consequences** of the human activities in the landscape. They are incorporated in various **regulations, laws, standards, plans, concepts, agreements, conventions** or other development documents.

Basic division and character of socio-economic phenomena are shown in Fig. 1.5 (more details see in Chap. 5).

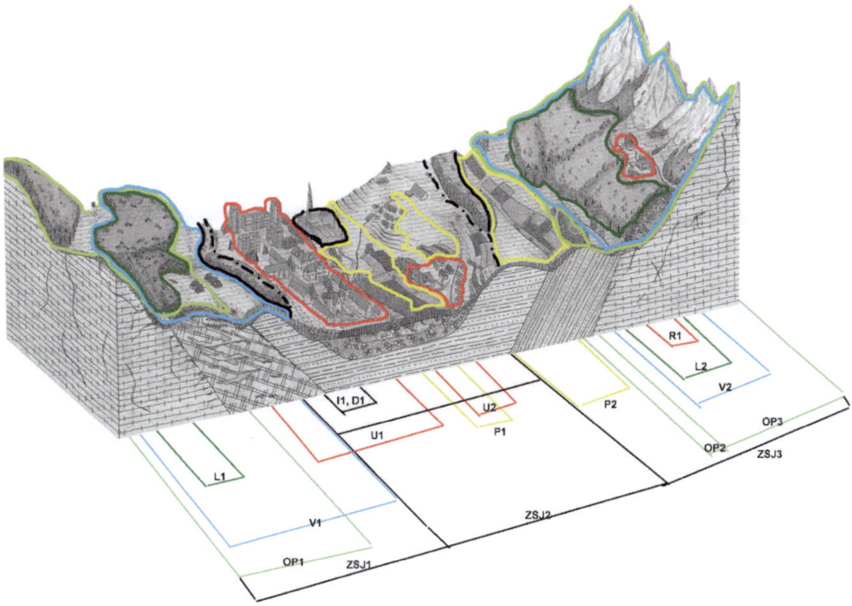

Fig. 1.5 Socio-economic factors in the landscape and their mutual overlaps. SEF linked to: conservation of nature—OP, protection of waters—V, soils—P, forests—L, urbanisation and recreation—**U, R**, production and technical objects—**I, D**, administrative division—**ZSJ**

1.4.3 Elements, Properties and Relations in Geosystems

As described above, on the basic distinguishing level, the elements of geosystems are the **components of the geographical sphere**. The **properties** of the elements are characterised by the **indicators** of these properties, which obtain certain **values** in the landscape.

> The following terms are to be clearly distinguished:
> **Geosystem/geocomplex**,
> **element (component) of the system**,
> **property** of an element,
> **indicator** of a property,
> **value** of the indicator.

The indicators may have the character of:

(a) **state variables**—real measurable values of particular indicators of the properties of elements (e.g. the depth of soils, amount of rainfall, flow rate, tree height, amount of biomass), which determine the current status of an element.

(b) **typological characteristics** (spatial subsystems with typological values e.g. soil type, climate type, physiognomic-ecological formation of plants). They are often expressed verbally, and provide a comprehensive description of numerous properties of a particular element. These characteristics are commonly projected on maps in form of spatial units.

It should be noted that it is impossible to lay down precise rules as to which elements should be characterised by state variables or typological characteristics. In general, it can be argued that state variables are suitable for large-scale landscape-ecological works as well as specialised studies, whereas typological characteristics seem to be preferred in informative and descriptive studies and in less detailed works on smaller scales. However, the final choice of the indicators depend on the given task and on the character of concrete territory.

Physical substance of the relations in geosystems is represented by **energy-material and information flows** between respective geosystem elements.

The relations, the energy-material and information flows in geosystems, which can be also called **processes**, can be determined by:

- '**measuring**' the values of the state variables of the indicators of the properties of those geosystem elements which enter the examined relationship—e.g. by taking measurement of the amount of rainfall and the amount of removed soil to determine the relation of rainfall and soil erosion. Such a method of determination is typical for specialised analytical geographical, environmental and other science disciplines, in which the evaluation of the relationship between selected elements represents the main object of research.
- by **comparison of the values** of state variables (typological characteristics) of one element with the value of state variables (typological characteristics) of the other element—e.g. characterisation of the relation between climate and vegetation by evaluation of plant communities and altitude. In this case, it is helpful that we know the results of these relations based on many years of specialised analytical studies and subsequent comparison of characteristics and synthetic assessments of the elements of geosystems. This is typical also for landscape ecology, which often utilises 'soft' systems with the sets of data commonly referred to as a 'fuzzy data set'.

In the landscape ecology also often used is the term '**autoregulatory mechanisms**'. Essentially, autoregulatory mechanisms govern energy-material-information flows, which, in space and time, maintain certain conditions in geosystems. It needs to be emphasised that the autoregulatory mechanisms are constantly at work, in any conditions, in primeval forests, deserts, heaps, sewers, regulated watercourses and even tarns. Man can change some indicator values of the properties of the elements, which can disrupt, accelerate or hamper the process of autoregulatory mechanism, but he cannot eradicate them, e.g. man can change the soil surface by heaping a tailings pile, but he cannot prevent the growth of pioneer plants on the pile. Even if he tried to exterminate them with another layer of tailing rock and thereby leave the

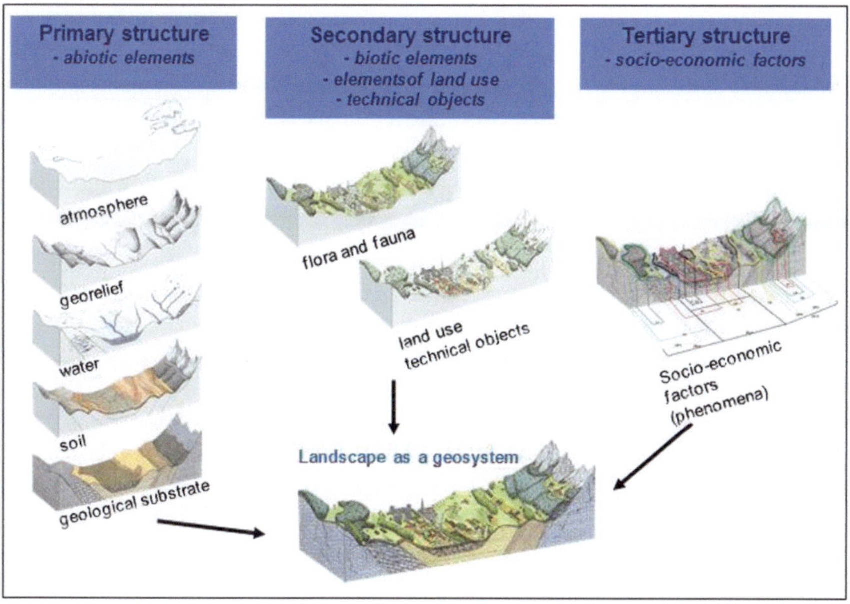

Fig. 1.6 The functional division of the structure of the landscape with regard to LANDEP

surface exposed, he cannot prevent erosion from happening on that surface. There are countless examples we could mention, but the point here is to remind that the process is an important aspect of the LANDEP methodology.

1.4.4 The Structure of the Landscape as a Geosystem According Its Function for LANDEP

A crucial aspect of the landscape as a geosystem is its structure (Armand 1949; Herz 1974; Haase 1980; Forman and Godron 1981; Csorba 1999, 2014; Brandt et al. 2013a, b). In relation to LANDEP, as well as to other plannings is crucial to define the material characteristics and functions of different part of landscape structure for the particular goals of the plannings. Accordingly, we developed the division of the landscape structure to three partial structures as the primary, secondary and tertiary landscape structure (Drdoš et al. 1995; Miklós and Izakovičová 1997; Miklós and Špinerová 2011). The listed studies divide the landscape structure, based on its evolution, physical character and relevance for landscape-ecological planning, into three substructures (Fig. 1.6):

1.4.4.1 Primary Landscape Structure

The ***primary landscape structure*** is a complex of those landscape elements and their relations, which, chronologically, constitute a primary and permanent basis for other landscape structures.

The elements of the primary landscape structure exhibit the following notable characteristics:

- they are physically **bound** to a particular place on earth;
- **man has not changed their functional principles** (e.g. water flows downhill, southern slopes receive more heat than the northern ones, prevailing wind directions are given, weathering of rocks cannot be stopped, etc.), but on the other hand;
- their reaction after their disruption is difficult to control;
- chemical, physical and structural substance is very difficult or even impossible to change (e.g. limestone is not granite, where mountain ranges are there are not the lowlands, cold climatic region is not a warm climatic region, etc.);
- these elements have undergone **fewer significant changes** in comparison with the secondary and tertiary landscape structure, which were strongly influenced, transformed or directly **created by man**.

The primary landscape structure composes especially of **abiotic elements of the geosystem**: geological substrate and subsoil, soils, then relief, water and air.

Some selected state variables and typological characteristics of the indices of the properties of elements of the primary landscape structure are shown in Tables 1.3a and b (more detailed characteristics see in Sect. 3.2).

1.4.4.2 Secondary Landscape Structure

The ***secondary landscape structure*** is a complex of physical elements **influenced, transformed and created** by man, which currently fill the **earth's surface**. They were created by human activity, by means of transformation of the primary landscape structure, so they appeared secondarily, later than the primary landscape structure (Ružičková and Ružička 1973; Miklós and Izakovičová 1997).

The elements of the secondary landscape structure exhibit the following notable characteristics:

- they are **physiognomic** elements of **land cover**,
- they are physically **bound** at a specific location on earth,
- they are **changeable** by exerting a certain amount of energy (e.g. forests into fields, fields into buildings, buildings into shrubberies, etc.).

The most representative characteristic of the secondary landscape structure is the land-use/land cover, which encompasses also the current vegetation and fauna, and the anthropogenic objects and materials.

Table 1.3 Chosen characteristics of the primary landscape structure used in LANDEP

State variables	[a]Typological characteristics
Geological substrate: rocks, weathered rocks, sediments	
Texture and structure	*Geological formations (geological maps)*
Resistance against weathering	*Quaternary formations*
Chemical structure (mineral strength)	*Types of rocks, lithological character*
Engineering geological substrate value	
Permeability and aquifer	
Subsurface water	
Resource efficiency	*Types of underground water according to permeability*
Chemistry and mineralization	*Types of mineralised underground water*
Depth of the underground water above sea level	
Depth of the level under the terrain	
Regime of underground water	
Soils	
Depth and skeletality	*Soil types and subtypes*
Texture (grain size)	
Structure, consistency, waterlogging by surface water	
Water, salt and thermic regimes	
Humus and carbonates content	
Chemistry and soil sorption	
Georelief	
(a) Topic (vertical) parameters of relief	*Morphotopes*
Coordinates	
Altitude and relative terrain heights	
Gradient of relief	
Horizontal and normal curvature	
Forms of relief	
Orientation of the relief to the cardinal points (aspect)	
(b) Choric (horizontal) parameters of relief	*Morphochores*
Gradient curves—direction of gravity	
The same direction of gradient—dividing lines	*Elementary gravitation areas, microcatchments*
Length of the gradient curve	
Shape and size of elementary gravitation areas	
Integration mode of elementary gravitation areas	
State variables	[a]Typological characteristics
(c) Topical-choric parameters of relief	*Morphotopochores*
Topographic location of morphotopes	*Morphogenetic types of relief*
Topographic location of relief types	*Orographic units*
Orographic position of relief types	*Natural-positional regions*
Vertical and horizontal segmentation of mesorelief	*Morphographic-morphometric types of the relief*
Total relief segmentation	*Other forms of relief types*
Relative elevation of morphotopes	
Relative elevation of microcatchments and depression	
Position and connection of microcatchments	*Types of microcatchments by location*
Neighborhood form of microcatchments, distance of microcatchment from the end of hydrological system	

(continued)

Table 1.3 (continued)

State variables	[a]Typological characteristics
Complex relief—surface water	
Runoff coefficient and specific surface runoff	
Integration of runoff (quantity)	
River basin area	
Spatial structure and shape of hydrological systems	
Hierarchical division of hydrological systems	*Individual river basins*
River flow and the amount of runoff	
Drain regime and other river regimes (thermal, ice drift, sediments regime)	*River types (by regime)*
Complex relief-climate	
(a) Meteorological indicators	
Basic: temperature, rainfall, humidity, pressure (averages, extremes, duration, regime)	*Climate regions and subregions*
Purpose indicators: temperature inversion	*Climate geographic types*
Purpose indicators: torrential rain, specific expense	*Agro-climatic areas*
Wind conditions (speed, direction, duration)	*Phenological-geographic types*
(b) Morphoclimatic indicators	
Insolation of the georelief (duration, quantities of incident sunlight)	*Morpho-climatopes*
Shading of the relief and Running the relief shadow	*Morpho-climachores*
Coefficient of deceleration Or acceleration of wind	*Morpho-climatopochores*
Shading of relief against different landscape elements	
Potential vegetation (as interpretation of the abiotic complex ABC)	
Theoretically predicted species composition of vegetation	*Geobotanical units*
Ecological and physiognomic characteristics of potential vegetation formation	*Phytogeographic areas*
Theoretically predicted species composition of fauna	*Regiotyps of fauna and zootesseras*
State variables of other elements of geosystem (climate, water, soil, geological substrate, relief)	*Typological units of other elements of geosystem (climate, water, soil, geological substrate, relief)*

[a]*Typological characteristics do not connect necessarily to the status variables in given row!*

So, the same element of secondary landscape structure can be characterised in terms of:

- **land-use forms** which created the **land cover** (physical forms of land use)—*physiognomic-functional perspective;*
- their **biotic** components (real flora and fauna)—*physiognomic-ecological perspective;*
- their **spatial structure**—*structural–spatial perspective*.

Some selected state variables and typological characteristics of the secondary landscape structure often used in the LANDEP are shown in Table 1.4 (more detailed characteristics see in Sect. 3.3).

All forms and manifestations of the secondary landscape structure—from of so-called cultural landscape, which landscape-ecological studies consider an ideal

Table 1.4 Chosen characteristics of the secondary landscape structure used in LANDEP

State variables	*Spatial subsystems*
Complex of real vegetation cover	
Species and its characteristics	
Physiognomical-ecological formation and its characteristics	***Physiognomical-ecological-functional-spatial units of real vegetation: mapping units of CLS)***
Species composition: floristic characteristics	
Grouping of species into communities: geobotanical characteristics	
Population characteristics: frequency, dynamics/density	
Ecological/site characteristics: humidity, soil response, trophic and climatic conditions	
Biostructural characteristics: life forms, shapes, cover, dimensions, vegetation layer, age, phenological phases, diversity, distribution of diaspore	
State of the system/state function: balance, synantrophy, hemoroby	
Production characteristics: feeding potential, honey production, therapeutic effect, plant toxicity	
Chorological characteristics: site size and its variation in time	
Ecosozological characteristic: rarity, vulnerability, endemits, relicts, protected species	
Complex of *biotopes*	
Physiognomic-ecological characteristics of biotopes of fauna	***Biotopes of fauna community***
Species composition and clusters within zoocenoses	
Population characteristics: abundance, diversity	
Characteristics of bioindicators	
Ecosozological characteristics	
Complex land use	
Land-use forms	***Elements of land use (land-use, land cover)***
Structural characteristics: size, shape of parcels	
Functional properties: position, availability, machinability	
Complex of urban-technical structures	
Functional-spatial characteristics: mode of the use of areas and objects	***Urban-technical elements***
Structural properties: shape, size, material, spatial expression	

state, to built-up industrial areas—came to existence by means of **land-use** activities. According to the economic–geographical terminology, CLS elements represent forms of land use (Ružička et al. 1978; Žigrai 1994; 1995). Other studies refer to the same elements as elements of land cover (Feranec and Oťahel 2001). Based on their biological–ecological content, these elements can be characterised as physiognomic-ecological formations—elements of **real vegetation** as well as **biotops** or **habitats** (Agger and Brandt 1988; Ružičková et al. 1996). The same elements of the secondary landscape structure can be then defined as a land cover element, actual vegetation cover, habitats of animal species, land use, technical-urban landscape structure.

A basic classification of mapping units of CLS still follows the original division of the secondary landscape structure, as proposed by Ružičková and Ružička (1973). The only difference is the level of detail which is determined by the objectives of a study.

The contemporary state of the secondary landscape structure is called **the current landscape structure (CLS)**. Its elements have the character of topical-spatial units of physiognomic-ecological formations (Miklós and Izakovičová 1997) and are used as **CLS mapping units**. Each planning procedure of ecologically optimal management and use of land, but also protection of nature and natural resources should result to the determination of a method of utilisation of each spot of the territory (Haber 2005, 2008; Štefunková 2011; Špulerová et al. 2013). So, we can state that the elements of SLS/CLS are the main operational units for planning and management procedures.

1.4.4.3 Tertiary Landscape Structure

The ***tertiary landscape structure*** is a complex of **non-material (intangible) elements** which are marked as **socio-economic factors/phenomena**. As explained in the Sect. 4.3, they express the **interests, manifestations and consequences** of **landscape-ecologically relevant** activities in the space. The elements of the tertiary landscape structure exhibit the following notable characteristics:

- they are intangible themselves, but are linked to tangible elements of the primary and secondary landscape structure,
- they need to be **landscape-ecologically relevant**, i.e. they have **spatial manifestation**, and thus are ‘mappable’.

As they are non-material, intangible (they have no physical representation), there is no reason to speak about their physical changeability, they physically do not exist. Nevertheless, they **considerably influence** the use of land in the past, as they do so in the present and will continue in the future.

The concrete elements of the tertiary landscape structure—the socio-economic factors (SEF) or socio-economic phenomena (as non-material entities)—are defined in acts, directives, regulations, norms, conventions, plans, conceptions, development documents. They are of the character of zones—e.g. conservation areas, belts—security and protective belts of industrial and transport objects, regions of nature resources protection. All these socio-economic factors are characterised in normative documents with very diverse kinds of restrictions, bans, but also supporting measures, allowances, even subsidies.

The basic division of socio-economic factors are shown in Table 1.5 (more detailed characteristics see in Sect. 3.4).

1.4.5 Interrelations of the Landscape Structures with LANDEP and Other Environmental Management

An important aspect of the division of the landscape into the above-mentioned structures is their mutual relation, especially with respect to **planning and projecting practice** (Verrasztó 1979). The decisive impact of these three structures on planning and management of the landscape, according to the logic of their characteristics should be as follows:

- as the primary landscape structure, PLS has immutable principles of operation, impossible and difficult to change properties, but easily changeable quality, **planning should primarily adapt utilisation of landscape to its characteristics, where possible not to change them**;
- the secondary/current landscape structure, SLS/CLS, is changeable by using an amount of energy. Planning can, therefore, **consider its changes**, but with respect to its quality, if possible according to the properties of the primary landscape structure;
- as the tertiary structure does not physically exist, in theory, it is the easiest to change. Therefore, it should be carefully **adapted to the characteristics of primary and secondary landscape**.

However, the last mentioned statement is a difficult issue. Naturally, the occurrence of the elements of the landscape structures is not incidental. As mentioned above, socio-economic phenomena SEF, although intangible, are linked to specific tangible elements, or specific combination of material elements of the primary or secondary landscape structure. Some SEF always co-occur with a certain element of the primary or secondary landscape structure; some occur only in specific cases. Several examples are given below: In case of the occurrence of a protected plant or animal species wherever, their area of occurrence is obligatory linked to a SEF—protected habitat. On the other hand, not each territory which exhibits a landscape structure similar to that of, e.g. the Little Carpathians of Great Fatra will automatically be declared as protected landscape area or a national park. Not each territory with rich water resources will be declared protected water management area, e.g. not the entire Žitný ostrov Island, rich in groundwater, is declared as CHVO Žitný Island, just a part. Not each bigger river basin is classified as a significant water-management basin. But, for example, each transport line, product line or energoduct has its own protections zone designated, and each farm has a sanitary zone and every quarry has its safety

Table 1.5 Typological characteristics of socio-economic factors/phenomena in the landscape

Main SEF groups	The content of SEF	Projection of SEF in maps
(a) SEF of nature conservation	Territorial nature protection	Protected area, protective zone
	Habitat of protected animals and plants	Protected area, protective zone
	Protection of species and individuals	Point (object), protective zone
(b) SEF of protection of natural resources	Protection and use of water resources	Protective zone/hygiene protection zone of an object (point), section
	Protection and use of natural healing resources	Protective zone/hygiene protection zone of an object (point), section
	Protection and use of forest resources	Site, USDF (unit of spatial division of forest), area
	Protection and use of soil resources	Parcel, site, administrative territory
	Protection and use of mineral resources	Site, point (object, borehole)
	Protection and use of cultural-historical resources	Structure, site, protective zone of site
(c) SEF of urban, residential and recreation areas	Urbanistic units and functional zones	Zone
	Recreational areas and zones	Zone, site, selected territory
(d) SEF of communication and transport	Protection and safety of transport structures	Protective/safety zones
	Protection and safety of energy/product pipelines	Protective/safety zone
(e) SEF of industrial and mining activities	Protection and safety of municipal facilities and sites (waste water treatment, waste management)	Protective/safety zone
	Protection and safety of storage facilities and sites	Protective/safety zone
	Protection and safety of production and mining facilities and sites	Protective/safety zone
(f) SEF of agricultural activities	Protection and safety of agricultural technical facilities and sites	Protective/safety zone
	Protection of health in range of facilities and sites of animal production	Hygienic zones

(continued)

Table 1.5 (continued)

Main SEF groups	The content of SEF	Projection of SEF in maps
(g) SEF of territorial units—administrative boundaries	State-administrative and regional units	Territorial boundary
	Sectorial territorial units	Territorial boundary
(h) SEF of deterioration of natural resources and the environment	Declared territory with polluted environment	Borders, areas, sections, points and sites

zone etc. Given the large amount of SEF and their possible combinations, it is not feasible to provide a comprehensive overview of this situation, and it is possible only in a particular territory.

An insight into the situation is provided by the Table 1.6a, b, c, d, e. Two columns on the left side contain the material elements and structures of the landscape, two columns on the right side the non-material regulations related to material elements and structures.

Unchangeable and changeable landscape structures—the theory versus the practice

From the point view of environmental care, nature conservation, maintaining ecologically optimal management and utilisation of land, as well as from the point of view of planning processes, **unchangeable and partially changeable landscape, the PLS and SLS/CLS**—the current landscape-ecological conditions—are the most critical structures because their disruption causes all ecological problems, whereas tertiary structure of the landscape TLS—the requirements of humans—**should adjust** to the primary and secondary landscape structure.

In practice, these principles are reflected in a different order:

- if something is defined by the tertiary structure (socio-economic phenomena), for example, by acts, regulations, plans, regardless of the correctness of their existence, for planning they are many times almost unchangeable obligatory principles because the authorities insist on these principles very strongly;
- the secondary structure is variable, so if it is not limited by any element of tertiary structure, it is only a slight obstacle to planning;
- in large areas of territory the primary structure is not limited by any regulations, e.g. relief, climate, etc., so planning only respects it if it conflicts with tertiary or secondary structures.

With a certain degree of simplification, it can be claimed that the SEF from the group of transportation, communal–technical activities, industrial and mining activities and agricultural activities are usually linked to protection, sanitary or buffer zones. SEF from the group of nature conservation as well as recreation and housing are not necessarily always linked to any elements of the primary or secondary landscape structure. SEF of administrative-territorial division is also not permanently

Table 1.6 Interrelation of socio-economic factor/phenomena SEF with the elements of the current landscape structure CLS

a. Green infrastructure

Elements of the CLS: **green infrastructure:**		*Tertiary landscape structure: socio-economic factors/phenomena SEF*	
Landscape as a whole	Specific landscape structure	*Protected landscape area*	***SK Territorial nature conservation***
		National park and buffer zone	
		Protected site and buffer zone	
		Natural reserve and buffer zone	
		Nature monument and buffer zone	
		Private protected area	
Mosaic of woods, grassland, wetlands	Mosaic of floodplains, groves Scrubs Lanes and line trees	*NATURA 2000: Sites of Community Importance/Protected Bird Area*	***Other systems of nature protection***
		Biotopes of Community Importance	
		Biosphere reserves of MAB UNESCO	
		UNESCO World Cultural and Natural Heritage Site	
		Biologically and aesthetically important area	
		Historical landscape structures	
		Threatened valuable landscapes	
	Wetlands	*Protected wetland ecosystems included in Ramsar convention*	
	Specific structures	*Biocentres, biocorridors, interacting elements*	***TSES Territorial system of ecological stability***
	Corresponding elements of CLS	*Protected plant*	***Species protection***
		Protected animal	
		Protected mineral	
		Protected fossil	
	Solitaires	*Protected tree*	
	Gardening areas	*Registered areas*	

(continued)

Table 1.6 (continued)

b. Forests, waters agricultural elements

Elements of the CLS: **forests, waters, agricultural elements**		*Tertiary landscape structure: socio-economic factors/phenomena SEF*	
Forests	Coniferous forests Broad-leaved forests Mixed forests Monocultures	*Commercial forest*	***Forests sources***
		Forest nurseries	
		Protection forests	
		Forest for specific purposes	
Waters	Wells, groundwater resources	*Protection zones of wells*	***Water sources***
		Protected groundwater area	
	Water flows, catchment areas	*Water catchment area*	
		Protection zones of water reservoirs	
	Water flows, lakes, water reservoirs	*Water suitable for bathing*	
		Water suitable for irrigation	
		Waters suitable for fish and for reproduction of original species	
		Fishing grounds, fish habitat protected area	
	Inundated alluviums, areas in-between dikes, polders	*Declared inundation territory*	
Other water elements	Thermal springs and other waters with specific properties	*Protection zones of natural healing and mineral waters*	***Healing water sources***
		Natural healing baths and their protection zones	
		Natural spa areas and their protection zones	
		Natural healing sources of gases and other emanations and their protection zones	
	Peatlands, poor fen	*Natural healing sources of peatlands, poor fens, muds and other soils and their protection zones*	
Arable land and other agricultural elements	All types of agriculture	*Parcels, registered culture in the real estate register*	***Soil sources***
	Arable land	*Bonited Soil-Ecological Unit (BPEJ)*	
		Drained and irrigated fields	
	Permanent grasslands	*Drained and irrigated fields*	
	Orchards, vineyards, hop gardens	*Registered permanent cultures*	
	Areas of greenhouses	*Investment of permanent character*	
	Gardening areas	*Registered areas*	

(continued)

Table 1.6 (continued)

c. Recreational and urbanised structures and objects

Elements of the CLS: **recreational and urbanised structures**		*Tertiary landscape structure: socio-economic factors/phenomena*	
Recreational areas and structures	Camping sites	*Declared zones of campgrounds*	***Recreation areas***
	Cottages area	*Cottage settlements*	
	Recreational centres	*Declared recreation areas, zones and centres of various level*	
	Spa and sanatoriums, hospital areas	*Zones of climatic spas*	***Healthcare zones***
		Protection zones of spas and sanatoriums	
		Protection zones of hospital grounds	
Urban areas	Parks and public greenery, cemeteries	*Recreation areas, rest zones*	***Urban functional zones***
	Sports areas	*Sport zones*	
	Family houses	*Residential zones for individual housing*	
	Block houses	*Residential zones for community housing*	
	Object and areas of public services	*Pedestrian zones, zones of public services*	
		Monument zones and reservations	***Cultural heritage***
		Cultural monument objects	
Communal–technical objects	Landfills	*Hygienic and protection zones (HPZ) of landfills*	***Technical zones***
	Ponds	*HPZ of ponds*	
	Incinerators	*HPZ of incinerators*	
	Wastewater treatment plants	*HPZ of wastewater treatment plants*	
	Stores	*HPZ of stores*	
Transport and telecommunication objects	Ports	*HPZ of ports*	***Transport zones***
	Airports	*HPZ of airports*	
		Safety zone of landing cones of airports	
	Railways	*Protection belts of railways*	
	Roads	*Protection belts of roads*	
	Electricity lines	*Protection belts of electricity lines*	
	Gas and oil pipelines	*Protection belts of gas and oil pipelines*	
	Telecommunication lines	*Protection belts of telecommunication lines*	

(continued)

Table 1.6 (continued)

d. Industrial and agricultural structures and objects

Elements of the CLS: **Industrial and agricultural structures**		*Tertiary landscape structure: socio-economic Factors/phenomena*	
Industrial and mining objects	Objects and areas related to mineral resources	*Natural rock structures for gas storage*	***Industrial and mining zones***
		Exclusive mineral deposits	
		Protected mineral´s deposit area	
		Mining areas	
		Protection and security zones (PSZ) of mining areas	
		Declared undermined zones	
		PSZ of quarries	
	Heaps, sedimental and deposit ponds	*PSZ of heaps and sedimental ponds*	
	Industrial objects and areas	*PSZ of industrial facilities*	
	Construction objects	*Protected (closed) construction sites*	
Agricultural technical objects	Agricultural—technical objects	*Protected areas of machineries*	***Agricultural and technical zones***
	Feedstocks	*Protected areas of feedstocks*	
	Livestock farms	*Hygienic zones of livestock farms*	
	Dumps of fertilizers	*Hygienic zones of dumps of fertilizers*	
	Objects of agrochemical enterprises	*Hygienic zones of agrochemical enterprises*	
Military objects	Field airports	*Protection zones of field airports*	***Military zones***
	Military objects and areas	*Security and guarded areas of military objects, shooting gallery*	
		Guarded training areas	
Exposed open substrate	Clay pits, borrow pits	*Protection zones of stone-pits, sandpits, gravel pits*	***Technical zones***
	Sand, gravel		
	Rocks and reefs		

(continued)

Table 1.6 (continued)

e. Deterioration of the environment			
Elements of the CLS: **related to deterioration**		*Tertiary landscape structure: socio-economic factors/phenomena*	
Landscape as a whole	All landscape elements	*Declared territories with the most affected environment*	***Deterioration of environment***
	Predominantly urbanised landscape	*Declared sources and areas with air pollution at different levels (above the limit)*	***Threat of natural resources***
		Areas with radiation, noise, vibration, odour	
		Registered sources of water pollution of different sizes	
		Segments of polluted water flows in different classes of pollution	
		Sensitive areas (Water Act)	
		Vulnerable areas (Water Act)	
	Predominantly agricultural landscape	*Areas with contaminated soils*	
		Areas with hardened soils	
		Areas with eroded soils	
		Areas with landslides	
	Predominantly forest landscape	*Forests affected by emissions*	
		Forests affected by pests	
	Whole landscape	*Territories threatened by avalanches*	
		Areas with seismicity	

linked to certain elements of the primary and secondary structure. SEF of deterioration of natural resources and the environment are linked only to an incidence of heightened concentration of pollution or other deterioration.

Definition of the landscape as a geosystem and its relation to legal status of LANDEP

Based on the definitions proposed by various authors, a landscape as a geosystem can be defined as a modification of the basic definition of system by Bertalanffy (1968) as: *a set of components (elements) of the geographical sphere and their interrelationships*. From the systemic, contentual and factual perspective, this definition can refer to material entities marked with different other terms as *landscape*, *geoecosystem, geocomplex, geographical complex*, but also other terms such as *territory, region, river basin, environment,* etc. whereas they cover the same particular material section of the geographical sphere, probably with slightly different borders. In other words,

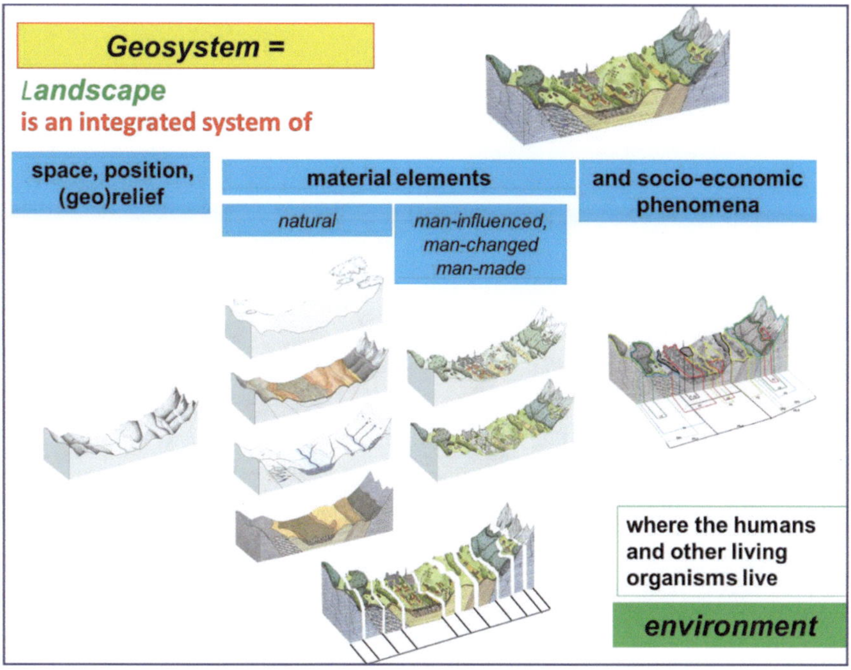

Fig. 1.7 The geosystem approach to the definition of landscape and environment

we can recognise on each particular material section of the geographical sphere at the same time geosystem, geocomplex, territory, region, river basin, etc. depending on our particular approach.

The definition of the landscape as a geosystem within the LANDEP methodology is an extended definition of basic system definition, as follows:

Landscape is a complex system of space, location, georelief and other mutually, functionally interconnected material natural elements and elements modified and created by a man*, in particular geological substrate and soil creating substratum, water bodies, soil, flora and fauna, artificial objects and the elements of utilisation of land, as well as their* ***relations determined by socio-economic phenomena in the landscape. The landscape represents the living environment of man and other living organisms***' (Miklós and Izakovičová 1997; Miklós and Hrnčiarová 2002).

The same wording of the definition of the landscape is presented in the Act No. 50/1976 Coll. on Territorial Planning and Building Order, in particular, in its amendment Act No. 237/2000 Coll. As explained in the Sect. 3.3, this definition is a base for the ecologically optimum spatial organisation and utilisation of the landscape, as such is the object of the landscape-ecological planning LANDEP. Last, but not least it means that the scientific definition of the landscape as a geosystem is legally confirmed also by law.

The definition can be graphically represented as follows (Fig. 1.7):

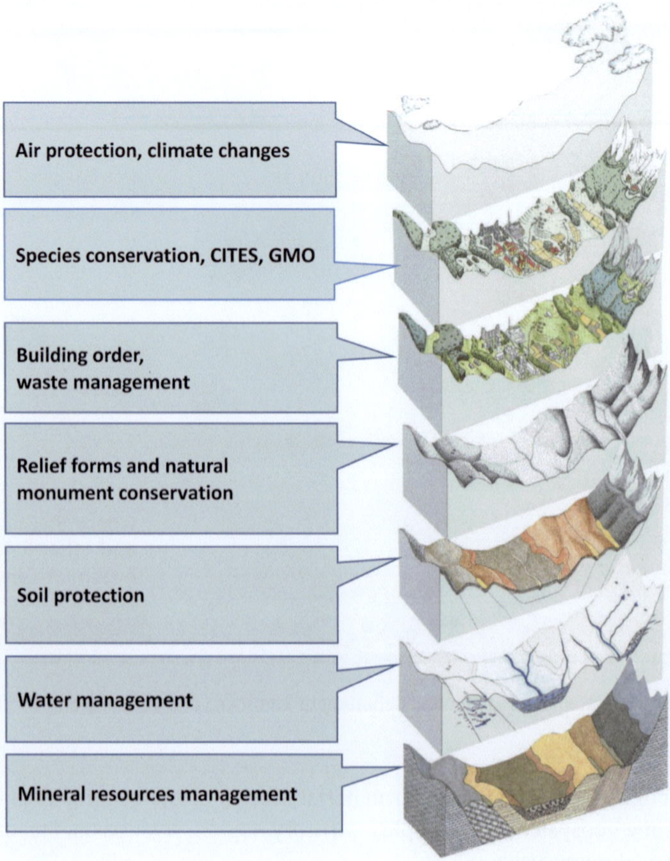

Fig. 1.8 The sectoral environmental management: protection of the components

Definition of the landscape as a geosystem and its relation to the environmental management

The definition of the geosystem as the environment serves actually also as a scheme for environmental management in the terms of state administration. The basic principles are:

(a) The topical model is traditionally the base for the **protection of individual components** of the environment. There are several state administration tools defined by acts for all components. The scheme of the protection of the components is show in Fig. 1.8. At the same time, this approach to the environmental management has a sectorial character, the responsibility for protection of different components of the environment are divided among ministries, as Ministry of environment, Ministry of agriculture, Ministry of transport and building,

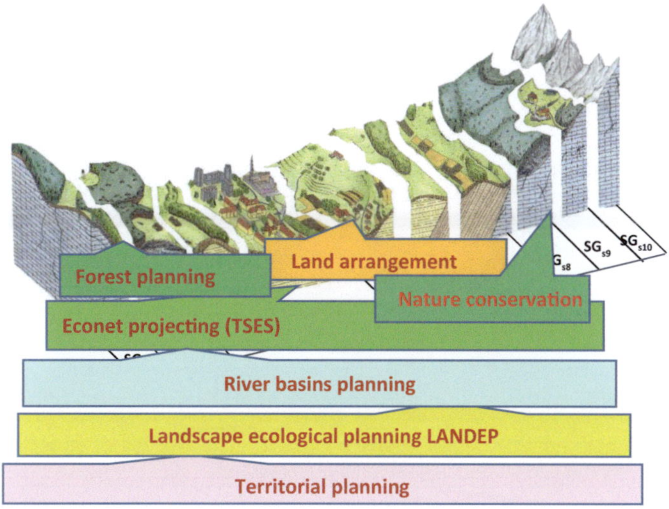

Fig. 1.9 The spatial planning and management policies: optimisation of the use of the spatial subsystems (complexes)

Ministry of health, Ministry of economy which bears different scope of responsibility, and sometimes they are in mutual contraversion.

(b) The choric model provides a scheme for the spatial/complex environmental management whereas management tools serves the legally defined spatial planning tools (Fig. 1.9). These tools are also in responsibility in different ministries (see Chap. 3), but the integrative character of all planning tools is given by the fact, they concern the space, where occurs all components of the geosystem, and, almost all sectoral activities. Therefore, this approach might be marked as **spatial/complex management of the environment**, which concern the protection of the spatial subsystems (complexes) using as tools the spatial planning procedures (Fig. 1.9).

(c) The model of **non-material** (**intangible**) socio-economic factors/phenomena displays the spatial distribution of the special **interests, manifestations and consequences** of the human activities mostly defined in law and other normative documents. Socio-economic factors are not bound to an area physically, but they influence all sectors, therefore, this model may explain the **horizontal politics**, as the state and progress of the environmental legislation, information system, and monitoring, economic tools, education and awareness rising, voluntary tools (see Fig. 1.10).

The landscape-ecological planning LANDEP concentrates on the ecologically optimum organisation and utilisation of the landscape with respect to utilitarian perspectives of the landscape structures as they represent:

- **conditions** for the life of mankind and all other living organisms;
- complex **natural resource** enabling life of humans and other organisms;

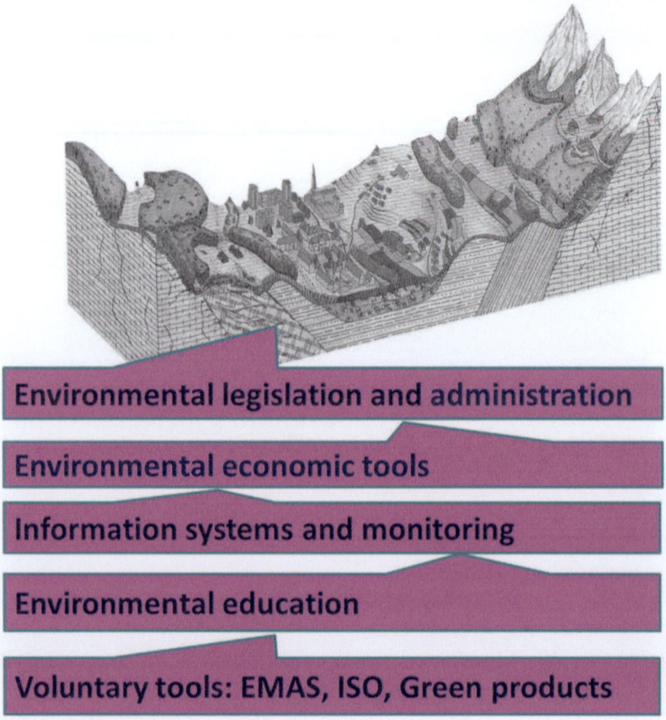

Fig. 1.10 The socio-economic model and the horizontal politics

- **potentials (of suitability):** prerequisites of utilisation for various purposes (Haase 1978; Miklós 1988; Tremboš 1994);
- provider of **ecosystem services** (Constanza 1997; Grunewald and Bastian 2015); but also as
- the stage of the **conflict of interests** of various activities (Izakovičová et al. 2010).

LANDEP is explicitly based on the understanding of the
landscape as a geosystem.
Accordingly, LANDEP applies
all **three models** and **three structures of the landscape**
in different stages of the methodical progress

The following chapters of this publication will discuss the methodical process of landscape-ecological planning LANDEP.

References

AGENDA 21 (1992) United Nations Conference on Environment and Development. Rio de Janeiro (United Nations), A/Conf. 151/ 4

Agger P, Brandt J (1988) Dynamics of small biotopes in Danish agricultural landscapes. Landscape Ecol 1(4):227–240

Ahrend Ch et al (1992) Spannungsfeld Landschaftsplannung. Zur Geschichte und Struktur eines heterogenen Faches. Schribri, Berlin

Alterman R (ed) (2001) National-level planning in democratic countries: an international comparison of city and regional policy-making. Liverpool University Press, Liverpool

Antrop M (2004) Landscape change and the urbanization process in Europe. Landscape and Urban Plann 67:9–26

Antrop M (2013) A brief history of landscape research. In: Howard P, Ian Thompson I, Waterton E (eds) The Routledge companion to landscape studies. Routledge, Oxon, New York, p 12–22

Antrop M, Van Eetvelde V (2017) Landscape perspectives. The holistic nature of landscape. Springer

Armand DL (1949) Zadachi geografii v svyazi s planom peredelki prirodi stepey. Voprosi geografii No. 13. Moskva

Armand DL (1952) Nauka o landschafte. Mysly, Moskva

Armand DL (1964) Logichnost geograficheskikh klassifikaciy i skhem rayonirovaniya. In: Razvitie i preobrazovanie geograficheskoy sredi. Moskva

Barančoková M, Krajčí J, Kollár J, Belčáková I (2010) Landscape ecology—methods, applications and interdisciplinary approach. Institute of Landscape Ecology Slovak Academy of Sciences, Bratislava

Barsch H, Saupe G et al (1993) Zur Integration landschaftsoekologischer und sociooekologischer Daten in gebietliche Planungen. Potsdamer Geographische Forschungen, Band 4. Universität Potsdam

Bastian O, Grunewald K, V. Khoroshev AV (2015) The significance of geosystem and landscape concepts for the assessment of ecosystem services: exemplified in a case study in Russia. Landscape Ecol 30(5):1145–1164. https://doi.org/10.1007/s10980-015-0200-x

Bastian O, Schreiber KF (1994) Analyse und ökologische Bewertung der Landschaft. Gustav Fischer Verlag Jena—Stuttgart

Belčáková I (2013a) Ochrana, tvorba a manažment krajiny. Trio Publishing, Bratislava

Belčáková I (2013b) Landscape Management. Theory and practice. Akva Rojal doo, Bački Petrovac

Belčáková I, Bianco del P, Finka M, Furdík J, Jaššo M, Petríková D, Wood B (2003) Aktuálne otázky priestorového plánovania a rozvoja. Road, Bratislava

von Bertalanffy KL (1950) An outline of general system theory. Br J Philos Sci 1:139–164

Bertalanffy KL von (1968) General System theory: Foundations, Development, Applications. New York: George Braziller. Penguin Books. revised edition 1976

Bezák P, Mederly P, Izakovičová Z, Špulerová J, Schleyer Ch (2017) Divergence and conflicts in landscape planning across spatial scales in Slovakia: an opportunity for an ecosystem services-based approach? Inter J Biodivers Sci Ecosyst Serv Manag 13(2):119–135

Brandt J (1999) Geography as 'landscape ecology. Dan J Geogr (Special Issue)1:21–32

Brandt J (2003) Multifunctional landscapes—perspectives for the future. J Environ Sci-China 15:187–192

Brandt J, Antrop M, Ramos IL, Padoa-Schioppa E, Porter J, Van Eetvelde V, Pinto-Correia T (2013a) How landscape ecology can promote the development of sustainable landscapes in Europe: the role of the European association for landscape ecology (IALE Europe) in the 21st century. Landscape Ecol 28(9):1641–1647

Brandt J, Christensen AA, Svenningsen SR, Holmes E (2013b) Landscape practise and key concepts for landscape sustainability. Landscape Ecol 28(6):1125–1132

Brandt J, Vejre H (2004) Multifunctional landscapes—motives, concepts and perceptions. In: Brandt J, Vejre H (ed) Multifunctional landscapes, vol 1, Theory, values and history, Witpress, p 3–31

Brandt J, Tress B, Tress G (Ed) (2000) Multifunctional landscapes: Interdisciplinary approaches to landscape research and management. In: Conference material for the international conference. Centre for landscape research, University of Roskilde, Denmark, 18–21 Oct 2000. Roskilde

Breuste J, Kozová M, Finka M (2009) European Landscapes in transformation: Challenges for landscape ecology and management European IALE Conference 2009, Salzburg (Austria), Bratislava (Slovakia)

Buček A, Lacina J, Lőw J (1986) Územní systémy ekologické stability krajiny. Životné prostredie 20(2):82–86

Buchwald K (1980) Umwelt - Mensch - Gesellschaft. Die Entstehung der Umweltproblematik. In: Handbuch fur Planung, Gestaltung und Schutz der Umwelt. Edited by Buchwald/Engelhardt BLV Verlagsgesellschaft. Miinchen, Bern, Wien

Canadian Institute of Planners (2013). https://web.archive.org/web/20130123140309/http://www.cip-icu.ca/web/la/en/default.asp

Chorley RJ, Kennedy BA (1971) Physical geography—a system approach. Prentice Hall Interantional Inc, London

Constanza R (1997) The value of the World´s ecosystem services and natural capital. Nature 387/1997

Cook EA, van Lier HN (eds) (1994) Landscape planning and ecological networks. Elsevier, Lausanne-New York-Oxford-Shannon-Tokyo

Csima P, Dublinszki-Boda B (ed) (2008) Tájökológiai kutatások. III. Tájökológiai Konferencia, Budapest

Csorba P (1987) A tájökológia időszerű kérdései az angol nyelvű szakirodalom alapján. Földrajzi Közlemények 35 (111.)1–2:74–80

Csorba P (1988) The ecogeographical pattern map of Tokaj "Great Hill" and its environs. In: Ruzicka, M., Hrnciarová, T., Miklós, L (ed) Spatial and functional relationships in landscape ecology. VIIIth international symposium on problems of landscape ecological research. IEBE CBES of SAS. IALE East European Region. Bratislava, vol 1, pp 273–284

Csorba P (1999) Tájszerkezet változások a Bodrogkeresztúri-félmedencében. Földrajzi. Közlemények 123(3–4):109–128

Csorba P (2014) A tájszerkezet földrajzi értelmezése. A geographical interpretation of landscape structure. 4D Tájépítészeti és kertművészeti folyóirat. J Landscape Archit Garden Art 36:32–39

Compendium of European planning systems (1997) Regional development studies report 28. European Commission. Office for Official Publications of the European Communities, Luxembourg

Convention on Biological Diversity (1992) United Nations Conference on Environment and Development (Earth Summit), Rio de Janeiro, July 3rd–12th

Cullingworth B, Nadin V (2006) Town and country planning in the UK. Fourteenth edition. Routledge, London

De Groot RS, Alkemade R, Braat L, Hein L, Willemen L (2010) Challenges in integrating the concept of ecosystem services and values in landscape planning, management and decision making. Ecol Complex 7(3):260–272

Demek J (1974) Systémová teorie a studium krajiny. Studia Geographica, 40, Brno, Geografický ústav ČSAV

Demek J (1978) The landscape as a geosystem. Geoforum 9(1):29–34

Deverre Ch, De Sainte Marie Ch (2008) L'écologisation de la politique agricole européenne. Verdissement ou refondation des systèmes agro-alimentaires? Revue d'Etudes en Agriculture et. Environment 89(4):83–104

Directive of the 2000/60/EC of the European Parliament and of the Council - Water Framework Directive

Directive 2007/2/EC of the European Parliament and of the Council of 14 March 2007 establishing an Infrastructure for Spatial Information in the European Community (INSPIRE)

Diviaková A (2010) Hodnotenie líniových formácií nelesnej drevinovej vegetácie pre potreby územných systémov ekologickej stability. VKÚ, a.s, Harmanec

Drdoš J (1967) Typizácia krajiny vo východnej časti Slovenského krasu a priľahlej časti Košickej kotliny. Biol. Práce, 12, 4, Vydavateľstvo SAV, Bratislava
Drdoš J (1992) Prírodné prostredie: zdroje – potenciály – únosnosť – hazardy – riziká. Veda Bratislava, Geografický časopis 44(1):30–39
Drdoš J (1995) Zamyslenie sa nad krajinným plánovaním. Životné prostredie 29(2):104–106
Drdoš J, Miklós L, Kozová M, Urbánek J (1995) Základy krajinného plánovania. Technická univerzita vo Zvolene
Drdoš J, Urbánek J, Mazúr E (1980) Landscape Syntheses and Their Role in Solving the Problems of Environment. Bratislava, Geografický časopis 32:119–129
Drdoš J (ed) (1983) Landscape synthesis. Geoecological Foundations of the complex landscape management. VEDA, Bratislava
European Landscape Convention (2000) Council of Europe, Florence
Europäische Raumordnungscharta (1983) Institut für Landes- und tadtentwicklungsforschung des Landes Nordrhein- Westfalen
European Spatial Development Perspective ESDP. Towards Balanced and Sustainable Development of the Territory of the EU (1999) European communities. Office for Official Publications of the European Communities, Luxembourg
Fabos JG (1979) Planning the total landscape, a guide to inteligent land use. Westview Press, Boulder CO
Fabos JG (1996) Greenways: the beginning of an international movement. Elsiever, Amsterdam
Feranec J, Oťahel J (2001) Krajinná pokrývka Slovenska (Land cover of Slovakia). Veda, SAV, Bratislava
Finka M, Petríková D, Jamečný Ľ (2008) Analýza právnych predpisov územného plánovania štátov EÚ ako podklad pre tvorbu územnoplánovacej legislatívy v SR (I. etapa - Nemecko, Rakúsko, Veľká Británia a Holandsko). URBION Bratislava, Centrum priestorových a technických štúdií, Bratislava
Foo K, Gallagher E, Bishop I, Kim AM (2015) Critical approaches to landscape visualization. Landscape and Urban Plann 142(Special Issue):244 pp
Forman RTT (1995) Land mosaics. The ecology of landscapes and regions. Cambridge University Press, Cambridge
Forman RTT, Godron M (1981) Patches and structural components for landscape ecology. Bioscience 31:733–740
Gercsák G (2011) GIS for the Ipoly River basin. In: Lin Jiun-Chuan (ed) Landscape conservation. Department of Geography, National Taiwan University, Taipei, pp 239–242
Golley FB, Bellot J (1991) Interactions of landscape ecology, planning and design. Landscape and Urban Plann 21:3–11
Grodzinski MD (2005) Piznannya landschaftu: misce i prostir (Understanding Landscape. Place and Space. In Ukrain). Kiev. U 2-x t. Vidavnicsho-poligrafichesky centr Kiivsky universitet. T.1
Grunewald K, Bastian O (eds) (2015) Ecosystem services—concept, methods and case studies. Springer-Verlag, Berlin, Heidelberg
Guidelines for Land Use Planning (1993) Food and agriculture organization of the United Nations, FAO Development Series 1, Rome
Haase G (1978) Zur Ableitung und Kennzeichnung von Naturpotentialen. Petermann's Geographische Mitteilungen, vol 122, pp 113–125
Haase G (1980) Izučenie topičeskich i choričeskich struktur, ich dinamiki i razvitija v landšaftnych sistemach. Structura, dinamika i razvitije landšaftov. Institut geografii AN SSSR, Moskva, pp 57–81
Haase G (1991) Naturraumerkundung und Landnutzung. Geochorologische Verfahren zur Analyse. Kartierung und Bewertung von Naturräumen. Beiträge zur Geographie. Berlin, 34, I und II
Haber W (1980) Entwicklung und Probleme der Kulturlandschaft im Spiegel ihrer Ökosysteme. Forsiarchiv 51(12):245–250

Haber W (1990) Using landscape ecology in planning and management. In: Zonneveld S, Forman RTT (eds) Changing landscapes on ecological perspective. Springer-Verlag, New York, pp 217–232
Haber W (2005) Landscape ecology and sustainability research. Z.f Geomorph. N.F. Suppl. Berlin, Stuttgart, vol 139, pp 1–18
Haber W (2008) Naturschutz in der Kulturlandschaft – ein Widerspruch in sich? NatSch_KuLa. Laufener Spezialbeiträge 1(08):15–25
Halada Ľ, Topercer J, Mederly P, Kartusek V (1995) Systém ekologickej kvality krajiny – ďalší prístup k manažmentu krajiny. Životné prostredie 29(5):271–273
Hall P (1992) Urban and regional planning, 3rd edn. Routledge, London, New York
Herz K (1974) Strukturprinzipien in der Landschaftssphäre. Ein Beitrag zur Methodologie der physischen Geographie. Geogr Ber 71:100–108
Hreško J, Bugár G, Petrovič F, Mederly P (2008) Endangering of abiotic complex as a basis for evaluation in the landscape-ecological plan of the Trenčín catastral area (Slovakia). Gospodarka surowcami mineralnymi 24(2):156–158
Hreško J, Mederly P, Petrovič F (2003) Landscape ecological research with support of GIS tools in prepration of landscape-ecological plan (model area of the Považská Bystrica city). Ekológia (Bratislava) 22(2):195–212
Hreško J et al (2015) Atlas archetypov krajiny Slovenska. Nitra, UKF
Hrnčiarová T, Miklós L et al (1997) Ekologická únosnosť: metodika a aplikácia na 3 benefičných územiach. I.-IV. Ústav krajinnej ekológie SAV a Ministerstvo životného prostredia SR, Bratislava
Huba M (1982) Štrnásť krokov na ceste za krajinným plánom. Geografický časopis 34:45–160
Hunziker M, Buchecker M, Hartig T (2007) Space and place—two aspects of the human-landscape relationship. In: Kienast F, Wildi O, Ghosh S (eds) A changing world. Challenges for landscape research, vol 8, Landscape Series. Springer, Dordrecht, pp 47 –62
Hynek A (2009) Prostorovosti: místa, krajiny, regiony. Acta Geographica Universitatis Comenianae No. 53:123–132
Hynek A (2010) Krajina: objekt, nebo konstrukt? In: Herber V (ed) Fyzickogeografický sborník 8, Fyzická geografie a kulturní krajina. MU, Brno, pp 138–142
Hynek A (2011a) Geografie, Geograficita – prostorovosti. In: Svobodová H (ed) (2011) Prostorovost: místa, území, krajiny, regiony, globiony. Sborník príspěvku z konference. GaREP, s.r.o. Brno, pp 6–50
Hynek A (2011b) Nazorova diverzita v chapani krajiny souvztažnost prostorovosti krajiny. In: Kolejka J et al (eds) Krajina Česka a Slovenska v sou.asnem vyzkumu. Masarykova univerzita, Pedagogicka fakulta, Brno, pp 12–46
Hynek A, Trnka P (1981) Topochory Dyjské částí Znojemska. Folia Facultatis scientiarum naturalium Universitatis Purkynianae Brunensis – Geographia, Brno, 22, 4:99
Implementation of the International Covenant on Economic, Social and Cultural Rights Second periodic reports submitted by States parties under articles 16 and 17 of the Covenant. Slovakia. 2009. UN Economic and Social Coucil, Committee on Economic, Social and Cultural Rights E/C.12/SVK/2
Isachenko AG (1981) Predstavlenije o geosisteme v sovremennoj fizičeskoj geografiji. Izv. VGO (Leningrad) 113(4), 297–306
Iverson L, Echeverria C, Nahuelhual L, Luque S (2014) Ecosystem services in changing landscapes: an introduction. Landsc Ecol 29:181–186
Izakovičová Z (1996) Landscape territorial system. Management and protection of environment. Faculty of Natural Sciences Comenius University, Bratislava, pp 285–292
Izakovičová Z, Mederly P, Petrovič F (2017) Long-term land use changes driven by urbanisation and their environmental effects (example of Trnava city, Slovakia). Sustainability: Open Access Journal [elektronický zdroj], 2017, vol 9, no 9, article no 1553
Izakovičová Z, Miklós L, Drdoš J (1997) Krajinnoekologické podmienky trvalo udržateľného rozvoja. Veda, Bratislava

Izakovičová Z, Moyzeová M (2007) Landscape-ecologically optimal spatial and functional organisation of landscape in the Tatry Biospeher Reserve of UNESCO. In: Guziová Z (ed) Priorities for conservation of biodiversity in biosphere reserves in changing conditions, Proceedings from the conference Stará Lesná, Tatry Biosphere Reserve, Slovakia, 2–6 June 2007. Institute of landscape ecology SAS, Bratislava, Slovak National Committee for UNESCO MAB Programme, pp 113–120

Izakovičová Z, Moyzeová M, Oszlányi J (2010) Problems in agricultural landscape management arising from conflicts of interest—a study in the Trnava region, Slovak Republic. Innovations in European rural landscapes. Springer, Heidelberg-Dordrecht-London-New York, pp 77–96

Izakovičová Z et al (2000) Metodické pokyny na vypracovanie projektov regionálnych ÚSES a miestnych ÚSES. Združenie krajiny 21, Bratislava

Izakovičová, Baránková Z, Dobrovodská M, Štefunková D, Babicová D, Moyzeová M, Petrovič F (2011a) Participation of local people on identifying the landscape values and future development in historical agricultural landscapes. Ekologia 30(1):216–228

Jančura P et al (1994) Manuál k metodickým pokynom pre spracovanie dokumentácií R-ÚSES. I.-V. diel.- SAŽP, odbor priestorového plánovania, Banská Bystrica

Johannesburg Declaration on Sustainable Development (2002) A/CONF.199/20. UN World Summit on Sustainable Development. Johannesburg, 26 August–4 September (2002)

Jones KB, Zurlini G, Kienast F, Petrosillo I, Edwards T, Wade TG, B-l Li, Zaccarelli N (2013) Informing landscape planning and design for sustaining ecosystem services from existing spatial patterns and knowledge. Landscape Ecol 28:1175–1192

Jongman RHG (1996) Research priorities: scientific concepts and criteria. In: Perspectives on ecological networks. European Centre for Nature Conservation, series Man and Nature, vol 1, chapter 14, pp 51–160

Jongman RHG, Pungetti G (2004) Ecological networks and greenways: concept, design, implementation. United Kingdom at the University Press, Cambridge, p 153

Kerényi A (2003) A földtudományok természetvédelmi feladatai az EU csatlakozás tükrében. Európai Úniós Csatlakozás és Földtudomány. Budapest, MTA Társadalomkutató Központ, pp 131–139

Kerényi A (2007) Tájvédelem. – Pedellus Tankönyvkiadó, Debrecen

Khoroshev A (2010) Multilevel analysis of landscape structure for land use decisions. In: Baranc˘okova´M, Krajčí J, Kollár J, Belčáková I (eds) Landscape ecology–methods, applications and interdisciplinary approach. Institute of Landscape Ecology, Slovak Academy of Sciences, Bratislava, pp 99–112

Khoroshev A, Koshcheeva A (2009) Landscape ecological approach to hierarchical spatial planning. Terra Spectra Plan Stud 1:3–11

Kienast F, Wildi O, Ghosh S (ed) (2007) A changing world. Challenges for landscape research, vol 8, Landscape Series. Springer, Dordrecht

Kocsis K, Agárdi N, Koczó F, Mezei G, Nemerkényi Zs, Szabó R (2016a) MNA alaptérképétől a Web GIS-ig. In: Balázs B (ed) Theory meets practice in GIS (in Hungarian). Debrecen: Debreceni Egyetemi Kiadó

Kocsis K, Agárdi N, Koczó F, Mezei GA, Nemerkényi Zs, Szabó R (2016b) Generalizálás géppel, vagy kézzel? - módszertani összehasonlítás Magyarország Nemzeti Atlasza példáján. In: Pajtókné Tari I, Tóth A (ed) Magyar Földrajzi Napok: Absztraktkötet : VIII. Magyar Földrajzi Konferencia. Budapest: Magyar Földrajzi Társaság, p 103

Kolektív autorov (2008) Katalóg tried objektov ZB GIS. Úrad geodézie, kartografie a katastra SR, Bratislava, Topografický ústav, Banská Bystrica

Koncepcia tvorby, aktualizácie a správy ZB GIS na roky 2006–2010. Úrad geodézie, kartografie a katastra SR (2006) Bratislava

Konečný M, Kubíček P, Duda P, Herman L, Kolář M, Kozel J, Mulíčková M, Řezník T, Stachoň Z, Štampach R, Štěrba Z (2012) Datové infrastruktury pro prostorově informační společnost. Masarykova univerzita, Brno

Koresawa A, Konvitz J (2001) Towards a new role for spatial planning. In: Organisation for economic co-operation and development: towards a new role for spatial planning. OECD, Paris

Kozová M, Bedrna Z (ed) (2003) Krajinnoekologické metódy v regionálnom environmentálnom hodnotení. Univerzita Komenského v Bratislave

Kozová M, Hrnčiarová T, Drdoš J et al (2007) Landscape ecology in Slovakia. Development, current state, and perspectives. monograph. Contribution of the Slovak Landscape Ecologists to the IALE World Congress 2007 and to the 25th Anniversary of IALE. Ministry of the Environment of the Slovak Republic, Bratislava, Slovak Association for Landscape Ecology—IALE-SK, 2007, CD ROM

Kozová M, Pauditšová E, Mišíková P (2010) Vývoj prístupov a metód podporujúcich rozvoj krajinného plánovania a aktuálne výzvy ovplyvňujúce jeho ďalšie smerovanie. In: Krajinné plánovanie. STU, Bratislava, pp 31–51

Krcho J (1968) Prírodná časť geosféry ako kybernetický systém a jeho vyjadrenia v mape 20(2):115–130 (Geografický časopis, Bratislava)

Krcho J (1974) Štruktúra a priestorová diferenciácia fyzicko -geografickej sféry ako kybernetického systému 26(2):132–162 (Geografický časopis, Bratislava)

Krcho J (1978) The spatial organisation of the physical-geographical sphere as a cybernetic system expressed by means of measure as entropy. Acta Fa Rer Nat Univ Comenianae, Geographica, 16, pp 57–147

Lamine C (2011) Transition pathways towards a robust ecologization of agriculture and the need for system redesign. Cases from organic farming and IPM. J Rural Stud 27(2):209–219

Land use planning applications (1991) Proceedings of the FAO expert consultation, Rome, 10–14 Dec 1990. World Soil Resources Reports 68. Rome, FAO

Langer H (1990) Ecological bricks for our common house of Europe. Politische Ökologie, 8, No. 21, Sonderheft 2, Muenchen, Verlag f. Politische Oekologie, 62 pp

Langevelde van F (1994) Conceptual integration of landscape planning and landscape ecology, with a focus on the Netherlands. In: Cook EA, van Lier HN (eds) (1994) Landscape planning and ecological networks. Elsevier, Lausanne-New York-Oxford-Shannon-Tokyo, pp 27–70

Leser H (1997) Landschaftsökologie. Ulmer, Stuttgart

Lőw J (1984) Zásady pro vymezování a navrhování územních systému ekologické stability v územne plánovací projekcii. Agroprojekt, Brno

Lőw J et al. (1984) Zásady pro vymezování a navrhování územních systému ekologické stability v územne plánovací projekcii. Agroprojekt, Brno

Löw J et al (1995) Rukověť projektanta místního ÚSES. Agroprojekt, Brno

Łowicki D, Mizgajski A (2013) Typology of physical-geographical regions in Poland in line with land-cover structure and its changes in the years 1990–2006. Geographia Polonica, Warszawa 86(3):255–266

Machar I, Kovář P (2010) Book of abstracts. In: International conference in landscape ecology “Landscape structures, functions and management: response to global ecological change”. Mendel University, Brno, CZ-IALE, Prague

Mander Ü, Helming K, Wiggering H (eds) (2006) Multifunctional land use: meeting future demands for landscape goods and services. Springer, Heidelberg, Berlin

Marsh WM (1997) Landscape planning: environmental applications. Wiley, New York

Mazúr E, Drdoš J, Urbánek J (1983) Krajinné syntézy - ich východiská a smerovanie. Geografický časopis 35:3–19

McHarg I (1969) Design with nature, Garden City. N.Y. Published for the American Museum of Natural History. Natural Hystory Press

Mederly P et al (2015) Integrated valuation of ecosystem service as a for landscape management and decision making at regional and local level (Case study Trnava, Slovakia). In: Ecosystem service for nature, people and prosperity, ecosystem services partnership

Mederly P, Krautschneider J, Halada Ľ et al (2012) Projekt pozemkových úprav v k.ú, Jasenica. Všeobecné zásady funkčního usporiadania územia. Regioplán, Nitra; Ekoped, Žilina

Mederly P, Hudeková Z (2005) Sustainable urban development in the Slovak Republic: proposal of set of indicators and their application in order to evaluate the sustainable development of cities. Bratislava, REC

Metodické pokyny na projektovanie územných systémov ekologickej stability (1993) Ministerstvo životného prostredia SR, Bratislava
Mičian Ľ (1982) A system approach to landscape and the sciences realizing it to the system of geographical sciences. In: VIth international symposium on problems of landscape ecological research. Oct 1982, Piešťany, Institute of Experimental Biology and Ecology SAS, Bratislava
Miklós L (1978) Náčrt biologického plánu povodia Gemerských Turcov. VEDA, Bratislava, Questiones Geobiologicae 21:127
Miklós L (1988) Space and position—scene of the origin of spatial ecological landscape problems. Bratislava, Ecology (CSSR) 7(4):381–395
Miklós L (1991) Zásady slovenskej ekologickej politiky. Životné prostredie (Bratislava) 25(4):174–178
Miklós L (1992) Protecting the Environment during Regional Development. In: Vasko T (ed) Problems of economic transition. Regional Development in Central and Eastern Europe. Chapter 5. AVEBURY, Aldershot—Brookfield USA—Hongkong—Singapore—Sydney, pp 51–62
Miklós L (2011a) The environmental policy of the Slovak Republic after the Velvet Revolution. Životné prostredie 45(6):295–300
Miklós L (2011b) the landscape in the international and slovak environmental politics. Životné prostredie 45(6):307–312
Miklós L et al (1985) Ekologický generel ČSSR. Časť SSR. I. etapa: Priestorová diferenciácia územia z ekologického hľadiska. Záverečná správa P16-121-402/01. ČSŽP Bratislava. ÚEBE CBEV SAV Bratislava, Stavoprojekt Banská Bystrica
Miklós L et al (1986) Ekologický plán VSN. Súbor grafických výstupov. Ekologická optimalizácia využívania VSN. IV. diel. ÚEBE SAV, Bratislava—Slovosivo
Miklós L, Diviaková A, Izakovičová Z (2011) Ekologické siete a územný systém ekologickej stability. Vydavateľstvo TÚ vo Zvolene
Miklós L, Hrnčiarová T (eds) (2002) Atlas krajiny Slovenskej republiky. 1. vydanie, MŽP SR Bratislava, SAŽP Banská Bystrica
Miklós L, Izakovičová Z (1997) Krajina ako geosystém. Veda SAV, Bratislava
Miklós L, Izakovičová Z et al (2006) Atlas reprezentatívnych geoekosystémov Slovenska, ÚKE SAV, MŽP SR, MŠ SR Bratislava
Miklós L, Kočická E, Diviaková A, Belaňová E (2011b) Integrated Landscape Management. Institutional Tools (in Slovak)..VKÚ, a.s., Harmanec
Miklós L, Kočická E, Kočický D, Diviaková A (2015) Geosystémy ako krajinnoekologická základňa pre integrovaný manažment krajiny. Vydavateľstvo Technickej Univerzity vo Zvolene
Miklós L, Špinerová A (2011) Krajinno-ekologické plánovanie LANDEP [The Landscape-ecological Planning LANDEP]. Harmanec, VKU
Minár J, Tremboš P (1995) Evaluation of the natural hazards in the landscape planning. Acta Environmentalica Univ Comenianae 4(5):211–222
Mizgajski A, Markuszewska I (ed) (2010) Implementation of landscape ecological knowledge in practice. The problems of landscape ecology, vol XXVIII. Polish Association for Landscape Ecology, Wydawnictwo Naukowe Adam Miczkiewicz University in Poznań
Moyzeová M (2008) Príklad revitalizácie poľnohospodárskej krajiny. In: Herber V (ed) Fyzickogeografický zborník 6. Fyzická geografie a trvalá udžitelnost. Příspevky z 25. výroční konference Fyzickogeografické sekce České geografické společnosti konané 31. a 31. ledna 2008 v Brne. Masarykova Univerzita v Brne, Brno, pp 34–38
Nassauer JI (2012) Landscape as medium and method for synthesis in urban ecological design. Landscape and Urban Plann 106:221–229
Nassauer JI (ed) (1997) Placing nature: culture and landscape ecology. Island Press, Washington, D.C
Nassauer JI, Opdam P (2008) Design in science: extending the landscape ecology paradigm. Landscape Ecol 23:633–644
Naveh Z, Lieberman A (1993) Landscape ecology—theory and application. Springer-Verlag, New York, Berlin, Heidelberg, London, Paris, Tokyo, Hong Kong, Barcelona, Budapest

Neef E (1967) Die theoretischen Grundlagen der Landschaftslehre. Haack, Gotha/Leipzig
Neef, E., Richter, H., Barsch, H., Haase, G. (1973). Beitrage zur Klärung der Terminologie in der Landschaftsforschung. Beitrag d. Institut d. Geographie and Geoökologie d. Akad. d Wiss. d. DDR. Annex to Proceedings of IIIrd International Symposium on the Landscape Ecological Research. Práce a materiály z biológie krajiny No. 20, UBK SAV Bratislava, 28 pp
Nesshöver C, Assmuth T, Irvine KN, Rusch GM, Waylen KA, Delbaere B, Haase D, Jones-Walters L, Keune H, Kovacs E, Krauze K, Külvik M, Rey F, van Dijk J, Vistad Odd I, Wilkinson M, Wittmer H (2017) The science, policy and practice of nature-based solutions: an interdisciplinary perspective. Sci Total Environ 579:1215–1227
Newman P, Thornley A (1996) Urban planning in Europe: International competition, national systems and planning projects. Routledge, London, New York
Opdam P, Foppen R, Vos C (2002) Bridging the gap between ecology and spatial planning in landscape ecology. Landscape Ecol 16(8):767–779
Oťaheľ J (1978) Fyzickogeografická regionalizácia Liptovskej kotliny. Questiones Geobiologicae. Problémy biológie krajiny, 20. Veda, vydavateľstvo SAV
Oťaheľ J (1994) Problémy environmentálneho plánovania (regionálne riešenia) In: Trvaloudržateľný rozvoj a krajinnoekologické plánovanie v európskych horských regiónoch, 1994. TU Zvolen, pp 166–174
Oťaheľ J, Lehotský M, Ira V (1997) Environmental planning: proposal for procedure (case studies. In Ekológia (Bratislava), 16(4):403–420
Petrovič F et al (2011) Environmentálne hľadiská trvalo udržateľného rozvoja. Nitra, UKF
Plány manažmentu povodňových rizík čiastkových povodí (2015) Ministerstvo životného prostredia, Bratislava
Plán manažmentu správneho územia povodia Dunaja (2015) Aktualizácia. Ministerstvo životného prostredia, Bratislava
Preobrazhensky VS, Minc AA (1973) Sootnoshenye ponyaty geosystema a ekosystema. In: Práce a materiály z biológie krajiny 20. In: Proceedings of IIIrd international symposium on the Landscape Ecological Research. ÚBK SAV, Bratislava
Preobrazhensky VS (1983) A system orientation of landscape research in geography and its present-day realization. In: Drdoš J (ed) Landscape synthesis. Geoecological Foundations of the Complex Landscape Management, VEDA, Bratislava, pp 31–36
Rao MV, Suresh Babu V, Suman Chandra S, Ravindra Chary G (2015) Integrated land use planning for sustainable agriculture and rural development. Apple Academic Press
Rega C (ed) (2014) Ladscape planning and rural development. Springer International Publishing, SpringerBriefs in Geography
Reitsmaa RF (1990) Functional classification of space. Aspects of site suitability assesment in a decission support environment. RR-90-2, IIASA, Laxenburg
Resolution of the European Parliament from 26th Oct 1990 on spatial planning. Strasbourg
Ružička M, Miklós L (1982) Metodické poznatky ekologického hodnotenia územia pre zónu a sídelný útvar (na príklade Rimavskej Soboty). Acta Ecologica, 9, 26, 74 pp (VEDA, Bratislava)
Ružička M, Miklós L (1982a) Metodické poznatky ekologického hodnotenia územia pre zónu a sídelný útvar (na príklade Rimavskej Soboty). VEDA, Bratislava, Acta Ecologica, 9(26):74
Ružička M, Miklós L (1982b) Landscape-Ecological Planning (LANDEP) in the Process of Territorial Planning. Ekológia (ČSSR), 1:297–312
Richling A, Solon J (1993) Ekologia krajobrazu. Wydawnictwo Naukowe PWN, Warszawa
Ružička M et al (2010) Krajinná ekológia ako vedecký základ krajinného plánovania. Krajinné plánovanie. Bratislava, STU p, pp 5–25
Ružicka M, Miklós L (1979) Teoretické a metodologické základy biologického plánovania krajiny. Záverecná správa úlohy VI-3-5/1. Bratislava, ÚEBE SAV
Ružička M, Miklós L (1990) Basic premises and methods in landscape-ecological planning and optimisation. In: Zonnenveld IS, Forman RTT (ed) Changing landscapes: an ecological perspectives. Springer Verlag, New York, pp 233–260

Ružička M, Ružičková H, Žigrai F (1978) Krajinné zložky, prvky a štruktúra v biologickom plánovaní krajiny In: Quaestiones Geobiologicae. Veda SAV, Bratislava, pp 9–10
Ružičková H, Halada Ľ, Jedlička L, Kalivodová E (eds) (1996) Biotopy Slovenska. Príručka k mapovaniu a katalóg biotopov. ÚKE SAV, Bratislava
Ružičková H, Ružička M (1973) Druhotná štruktúra krajiny ako kritérium biologickej rovnováhy. Quaestiones geobiologicae 12:23–62
Schilleci F, Todaro V, Lotta F (2017) Connected lands: new perspectives on ecological networks planning. Springer
Schmidt C (1993) On Economization and Ecologization as Civilizing Processes. Environ Values 2(1):33–46
Schreiber KF (1980) Brachflachen in der Kulturlandschaft. Daten und Dokumente zum Umweltschutz, Sonderreihe Umwelttaguug 'Okologische Probleme in Agrarlandschaften'. Univ. Hohenheim 30:61–93
Silberstein MA, Maser C (2013) Land-use planning for sustainable development, 2nd edn. CRC Press
Sklenička P (2003) Základy krajinného plánování. Vyd. Naděžda Skleničková, Praha
Snacken F, Antrop M (1983) Structure and dynamics of landscape systems. In: DRDOŠ J (ed) Landscape synthesis: Geoecological foundations of the complex landscape management. VEDA, Bratislava, pp 10–30
Sochava VB (1977) Vvedenje v učenije o geosystemach Nauka, Novosibirsk
Spangenberg JH (2016) Hot air or comprehensive progress? a critical assessment of the SDGs. Sustainable development. Wiley Online Library. https://doi.org/10.10002/sd.1657
Spatial Planning: Key Instrument for Development and Effective Governance with Special Reference to Countries in Transition (2008) UN publication, ECE/HBP/146. Economic Comission for Europe, Geneva
Špinerová A (2011) Prvky využitia zeme ako regulátor zón dynamiky svahu v modelovom území Ilijského potoka. Doktorská dizertačná práca, TU Zvolen
Spitzer H (1995) Einführung in die räumliche Planung. UTB, Stuttgart
Špulerová J, Dobrovodská M, Izakovičová Z, Kenderessy P, Petrovič F, Štefunková D (2013) Developing a strategy for the protection of traditional agricultural landscapes based on a complex landscape-ecological evaluation (the case of a mountain landscape in Slovakia). Moravian Geogr Rep 21(4):15–26
Štefunková D (2011) Vybrané výsledky výskumu starej vinohradníckej krajiny Svätého Jura. In: Ekologické štúdie: Vedecký časopis venovaný aktuálnym problémom ekológie, krajinnej ekológie a príbuzných vedných disciplín 2(1–2):126–131
Štefunková D, Hanušin J (2015) Analysis of the spatial and temporal distribution of selected landscape diversity indexes in detailed scale (Example Of Viticultural Landscape Svätý Jur). In: Halada Ľ, Bača A, Boltižiar M (eds) (2015) Landscape and landscape ecology. Proceedings of the 17th international symposium on landscape ecology. Institute of Landscape Ecology, Slovak Academy of Sciences, Bratislava, pp 185–191
Szabó SZ, Csorba P, Szilassi P (2012) Tools for landscape ecological planning—scale and aggregation sensitivity of the contagion type landscape metrics indices. Carpathian J Earth and Environ Sci 7(3):127–136
Szabó SZ, Csorba P, Varga K (2008) Landscape indices and landuse—tools for landscape managment. In: Plit J, Andreychouk V (ed) Methods of landscape research. Dissertations, Commission of Cultural Landscape 8:7–20, ISSN 1896-1460, Sosnowiec, Poland
The future we want (2012) Resolution adopted by the General Assembly on 27 July 2012. A/RES/66/288. UN General Assembly
The Pan-European Biological and Landscape Diversity Strategy (1995) Council of Europe, UNEP, ECNC, 50 pp
Turner, MG (1990) Spatial and temporal analysis of the landscape patterns. Landscape Ecology 4(1):21–30
Transforming our World: The 2030 Agenda for Sustainable Development (2015) United Nations

Tremboš P (1993) Potenciál krajiny, jeho hodnotenie a využitie v územno-plánovacej praxi. Životné prostredie 27:41–43
Tremboš P (1994) Krajinnoekologické podmienky a regionálny rozvoj. In: Trvaloudržateľný rozvoj a krajinnoekologické plánovanie v európskych horských regiónoch. TU vo Zvolene, pp 163–166
Tress B, Tress G (2001) Capitalising on multiplicity: a transdisciplinary systems approach to landscape research. Landscape and Urban Plann 57:143–157
Tress B, Tress G (2002) Disciplinary and meta-disciplinary approaches in landscape ecology. In: Bastian O, Steinhardt U (eds) Development and perspectives of landscape ecology. Kluwer Academic Publishers, Dordrecht, pp 25–37
Tsybuliak AG (2015) Ecologizing parameters of the world economy. Eur Sci Rev 1–3:163–165
United Nations Millennium Declaration (2000). Resolution adopted by the General Assembly. A/55/L.2
Verheye W (ed) (2004) Land use, land cover and soil sciences. Encyclopedia of Life Support Systems (EOLSS). UNESCO-EOLSS Publishers, Oxford, UK
Veteikis D, Kavaliauskas P, Skorupskas R (2015) Assessing the optimality of landscape structure in a landscape plan (a Lithuanian example) In: Halada Ľ, Bača A, Boltižiar M (eds) Landscape and landscape ecology. Proceedings of the 17th international symposium on landscape ecology. Institute of Landscape Ecology, Slovak Academy of Sciences, Bratislava, pp 348–358
Verrasztó Z (1979) Land formation and the geological aspects of environmental protection. Symposium changes of the geological environment under the influence of man´s activity. IAEG National group, Krakow-SandomierzBelchatow-Plock-Warszawa, pp 135–141
Verrasztó Z (2017) Holisztikus környezeti modellezés kartográfiai, GIS és távérzékelési módszerek alkalmazásával. Remote Sens 7(1):490–516
Vodný plán Slovenskej republiky (2009) Plán manažmentu správneho územia povodia Dunaja. Plán manažmentu správneho územia povodia Visly. MŽP SR. Eds. VÚVH, SAŽP, Vydavateľstvo SAŽP
Wehrmann B (ed) (2012) Land use planning. Concept, tools and applications. BMZ. GIZ Deutsche Gesslschsft fur internationale Zusamenarbeit GmBH on behalf of the Federal Ministry for Economic Cooperation and Development BMZ. Bonn-Eschborn
Williams RH (1996) European Union spatial policy and planning. Paul Chapman Publishing, London
Wrbka T, Erb KH, Schulz NB, Peterseil J, Hahn CO, Haberl H (2004) Linking pattern and process in cultural landscapes. An empirical study based on spatially explicit indicators. Land Use Policy 21:289–306
Wu J (2013) Key concepts and research topics in landscape ecology revisited: 30 years after the Allerton Park workshop. Landscape Ecol 28:1–11
Young A (1986) Land evaluation and diagnosis and design: towards a reconciliation of procedures. Soil Surv Land Eval 5:61–76
Young A (1993) Guidelines for land use planning. Food and Agriculture Organization of the United Nations, Rome, Italy. FAO Development Series 1
Zonneveld IS (1981) Land(scape) ecology, a science or a state of mind. In: Tjallingii SP, de Veer AA (ed) Perspectives in landscape ecology. Proceedings of the international congress organized by the Netherlands society for landscape ecology, Veldhoven 1981, pp 9–15; Wageningen
Zonneveld IS (1995) Land ecology: an introduction to landscape ecology as a base for land evaluation, land management and conservation. SPB Publishing, Amsterdam
Zonneveld IS, Forman RTT (eds) (1990) Changing landscapes: an ecological perspectives. Springer Verlag, New York
Žigrai F (1994) Landschaftsoekologische Planung als Instrument der nachhaltigen Entwicklung in Gebirgsregionen. In: Trvaloudržateľný rozvoj a krajinnoekologické plánovanie v európskych horských regiónoch. TU Zvolen, pp 135–139
Žigrai F (1995) Integračný význam štúdia využitia zeme v geografii a ekológii na príklade modelového územia Lúčky pri Liptove. In: Drgoňa V (ed) Geographical studies 4. University of Education, Nitra

Chapter 2
Landscape-ecological Planning, LANDEP—A Tool for the Ecologisation of Spatial Planning Processes

Abstract The basic principle of the landscape-ecological foundations of the environmental care is the integration of landscape-ecological ideas into the current planning processes—especially to the territorial planning, land-use, agricultural and forest management planning, and projecting. This integration is referred to as **ecologisation of landscape management**, more precisely: ecologisation of spatial organisation, utilisation and protection of landscape. These ecologisations have two main aspects: confrontation of societal demands for territorial development with the conditions of the landscape, which constitute limiting, restricting or promoting development regulations, and subsequent harmonisation of optimal arrangement of various activities in the concrete territory. LANDEP recognises these two basic stages of its methodology, as:

- **spatial–organisational**—proposal for ecologically optimal organisation, utilisation and protection of the landscape—the output has a character of management of 'space';
- **functional–technological**—proposal for relevant measures to ensure ecologically optimal "technology" of the performance of various activities within the areas delineated in the spatial-organisational step. This problem circle belongs partially to the competencies of respective sectors.

The main methodical principles in one sentence can be expressed as the confrontation and harmonisation of **all properties of a landscape**, with the **entire complex of demanded social activities** for the **entire concerned territory**. The methodology and methods in LANDEP can theoretically be derived into elementary steps from reverse analysis from final goals to start as follows:

- The ultimate goal of LANDEP is the ecologically optimal functional organisation of the landscape for human activities in the form of proposals—the step marked as **propositions**.

L. Miklós and A. Špinerová, *Landscape-ecological Planning LANDEP*,
https://doi.org/10.1007/978-3-319-94021-2_2

- The proposals need to determine the suitability of each element of the landscape for different activities from different perspectives—the step marked as **evaluation**.
- The evaluation's relevant information about the landscape—special-purpose criteria—interpreted indicators of landscape features derived from the basic data synthesised in geocomplexes. This step is called **interpretation**.
- The interpretations need to describe the landscape features comprehensively—in types and regions of landscape-ecological complexes, created within the process called as landscape-ecological **synthesis**.
- For the purpose of the synthesis, we must obtain the necessary information about the landscape properties. The process of their acquisition, modification and homogenization are represented by the step named as landscape-ecological **analysis**.

Obviously, the real landscape-ecological planning proceeds in opposite order as analyses–syntheses–interpretations–evaluations–propositions.

Keywords Ecologisation · Spatial organisation · Methodical aspects
Process of LANDEP

Spatial planning processes building on the landscape-ecological foundations are considered the essential tool of environmental care. The basic principle is the **integration of landscape-ecological ideas** into the current planning processes—especially territorial planning, land-use, agricultural and forest management planning and projecting.

This is referred to as **ecologisation of landscape management**, more precisely: ecologisation of spatial organisation, utilisation and protection of landscape (Miklós 1991, 2011a, b; Deverre and De Sainte Marie 2008). The process covers a complex of activities referred to also by other terms such as landscape maintenance, development and conservation of landscape, planning of the ecological management in the landscape and other similar.

2.1 Ecologically Optimal Spatial Organisation, Utilisation and Protection of the Landscape

The methodical principle of optimal spatial organisation is as follows:

- a confrontation of societal demands for territorial development with the conditions (properties) of the landscape, which, in this sense, constitute limiting, restricting or promoting development regulations,
- subsequent harmonisation and proposal of optimal arrangement of various activities within a territory, in particular, the organisation and utilisation of the landscape.

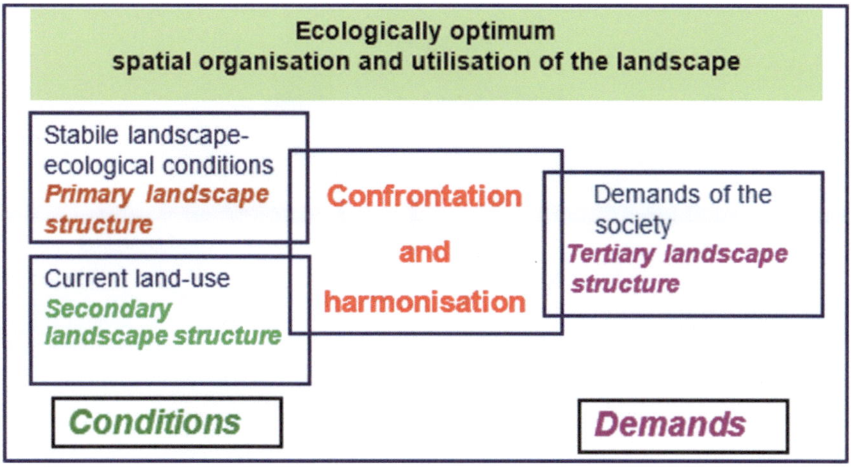

Such optimisation seeks to answer a triad of questions as

what, **where and how**

to manage the landscape, to maximise the compliance with the given conditions.

The landscape-planning processes primarily aim to answer the question **WHERE**? (Fig. 2.1).

Change or Preserve the Current Landscape Structure?

The fundamental question man has always asked is whether he likes or does not like the current landscape structure. Man lives at certain time in the landscape with a certain structure—**the current landscape structure** composed of **individual elements**—forest, fields, water bodies, residential areas and industrial parks and transportation areas. Such a structure, although temporarily satisfactory, becomes increasingly inadequate and forces man to institute changes to improve it, while altering the way the land is utilised, the boundaries of land-use elements and the current landscape structure.

How does Man React to a Particular Situation?

When the current landscape structure is, at a given point in time, assessed as **unsatisfactory**—which has been the case almost throughout the whole history of mankind—mankind strives to change the structure. Such changes have been initiated especially during agrarian revolutions, when man deforested large areas, or during Industrial Revolutions, when significant large-scale and large-volume technical structures were placed in the landscape. The question is if the changes have been made spontaneously or wisely. The dilemma has run through the entire history of humanity, and recently, it was formed into a paradigm called **sustainable development**.

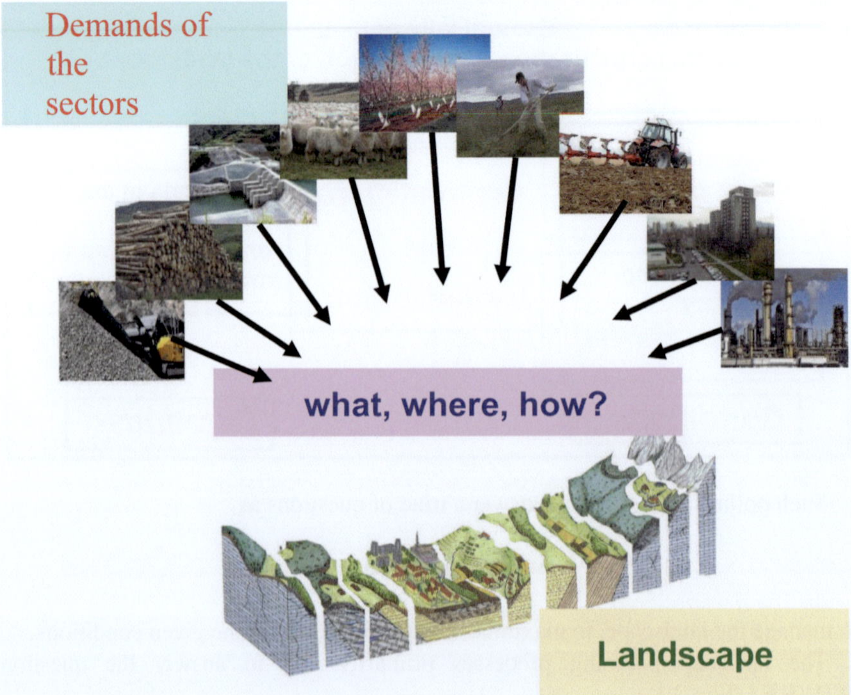

Fig. 2.1 Ecologically optimum spatial organisation and utilisation of the landscape

If a given structure is satisfactory, or mankind even "likes" it—they will prevent changes and seek to protect it. These represent predominantly efforts of nature and landscape **conservationists**. In this case, any extreme ideas need to be regulated by the principles of sustainable development.

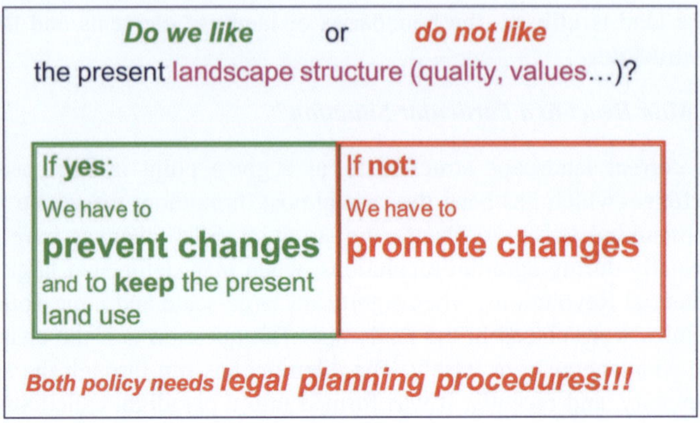

So, the LANDEP process also investigates this seemingly simple questions, namely whether the current arrangement and utilisation of the landscape, from the landscape-ecological perspective, is

suitable or **unsuitable**,

accordingly

if interpreted as unsuitable, **how to change it**?

if interpreted as suitable, **how to protect or conservate it**?

A rational social approach to this issue can be formulated as an effort to **achieve optimal change** or **maintain the current landscape structure**. It can even be understood as a constant historical "battle" of a man against landscape structures. However, such changes—intentional or unintentional, planned or not—affect the functioning of the **primary landscape structure** and trigger more or less adverse reactions. These simple facts need to be realised in case of each landscape-ecological conception, from a decision about an optimal way to utilise a certain area such as arable land, through addressing the issues of biodiversity conservation, balance and stability, to the currently popular objectives of evaluation of landscape character, historic landscape structures and the aesthetics of the landscape (Fig. 2.2).

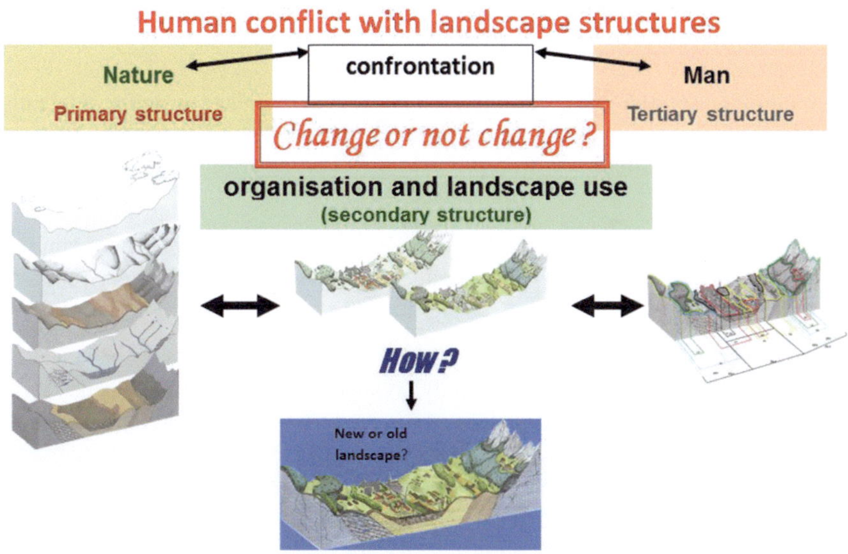

Fig. 2.2 Change or prevent—Conflict of man and landscape structures

2.2 The Main Principles and Methodological Aspects of LANDEP

The methodology of landscape-ecological planning represents a systematic procedure that can be considered a basis for ecologisation of spatial planning processes LANDEP (Ružička and Miklós 1982a, b, 1990).

The main objective of LANDEP is to prepare such landscape-ecological materials, which may serve as

a fundamental document for other spatial plannings

as well as a document facilitating decision-making processes in spatial planning especially for territorial plans and agricultural land arrangement projecting.

To ensure that the actions in the landscape comply with the conditions of current, but also primary landscape structure, and to mitigate the impact of any changes to utilisation of land, rational management needs to be implemented. A crucial component of all, even the most complex conceptions, is always the same: the spatial organisation, and the way of utilisation of areas.

LANDEP recognises these two aspects in its methodology as two basic stages of the process, as follows:

- **Spatial–organisational**—proposal for ecologically optimal organisation, utilisation and protection of the landscape—the output has a character of management of "space";
- **functional–technological**—proposal for relevant measures to ensure ecologically optimal "technology" of the performance of various activities within the areas delineated in the first step. This problem area partially comes under a sectoral competency.

The methodological nature of the process of **ecologically optimal spatial arrangement and utilisation of landscape** is likewise simple and has two major aspects:

- **characteristics of landscape properties**: the territory of interest is characterised by its properties, which are synthetically organised into quasi-homogeneous areas with clearly defined indicator values of the properties. In the LANDEP methodology, this step is elaborated in analyses and syntheses and its output represents a delineation and description of abiotic, biotic and **landscape-ecological complexes**.
- confrontation of the above-mentioned landscape properties with the demands man has within the concerned territory and subsequent decision-making and proposal for **changing or maintaining** the current state. This step is elaborated in the process of interpretations, evaluations and propositions.

Due to its relative simplicity, the whole process, especially for small areas with one or two projected activities, can take place quite quickly employing just plain horse sense, but a multistage professional procedure needs to be applied when performing confrontation, decision-making or harmonisation for larger areas. Insufficient application of such procedures leads to conflicts of interests and causes other problems in the landscape arising from inadequate spatial organisation and use of land.

The results of this process may have three major levels, which are as follows:

- determining suitable, conditionally suitable and unsuitable areas for utilisation for selected activities—defining **restrictions** and **limitations** on such utilisation. This is the most desired result of LANDEP. It is a multifaceted process as there are a lot of activities which might require landscape-ecological plans—from the arrangement of TSES elements to optimisation of recreational utilisation of interest areas.
- determining the most ecologically appropriate use of each area. The result of this line of decisions is the so-called **functional typisation of territory**. From the landscape-ecological point of view, this is the main output, which should constitute a basis for all other planning processes. However, this result is also problematic and less demanded as it is in conflict with the expectations and requirements of owners and users of land.
- determining of the most ecologically appropriate function of larger territorial units (zones, subregions, regions, etc.) in terms of their broader territorial relations. The result of this is the so-called **functional regionalisation of a territory**. This result should facilitate the optimisation of functional zones in the territorial plans of municipality, although currently they are required only very sporadically. It is caused mainly by the current practice of spatial planning in the country.

Based on the aforementioned theoretical and methodological principles, LANDEP can be defined as follows:

> LANDEP is a systematically organised special-purpose complex of applied landscape-ecological methods of analyses, syntheses, interpretations, landscape evaluations that result in determination of landscape-ecologically suitable, limited and unsuitable activities within an interest territory and in appropriate organisation of social activities in the landscape and application of relevant measures to ensure the functionality of these activities. This process is called optimisation of spatial organisation, utilisation and protection of the landscape.

The crucial methodological approach of landscape-ecological planning LANDEP to the object of planning, i.e. to the landscape is a geosystem approach as presented in Chap. 4, which interprets **the landscape as a geosystem**.

The LANDEP methodology is a strictly defined logical procedure, but, at the same time, it is an open system. The specific content of its methodological steps in particular works varies according to the following:

- the nature of the concrete task
- the required hierarchical level, scale of projection and the character of a given territory.

Currently, LANDEP focuses on the following types of tasks:

- Determination of limits and restrictions on land use. It is a procedure frequently utilised for planning and other management activities.
- The ecologically optimal organisation and utilisation of the landscape with the proposal for optimal functional division of a territory. This is the main direction of LANDEP methodology, defined in the Act No. 50/1976 Coll. According to its amendment 237/2000 Coll., especially for the needs of territorial planning and agricultural land arrangement.
- Territorial systems of ecological stability TSES. From the systemic point of view, the design of TSES is often a part of complex landscape-ecological plans.
- Ecological carrying capacity ECC. From the systemic point of view, the indicators of the ecological carrying capacity represent a comprehensive criterion for decision-making processes in LANDEP.
- Resolution of conflicts of interests as a basis for rational utilisation of natural resources and nature conservation.
- Resolution of special issues—environmental problems, greenery in the landscape, human impact on the landscape, erosion problems, soil problems, etc.

The content of a question "where and how to manage" will be dealt with differently at the state level which stresses the development of particular economic sectors in large regions, and at the level of an agricultural holding which addresses the issue of the appropriateness of crops for a particular stretch of land. Another decisive issue is the character of the landscape on the territory under planning.

A complex approach to the LANDEP methodology can be expressed by the following three essential aspects:

- aspect of landscape-ecological complexity (consideration of all three landscape structures, see Chaps. 1.4.4 and 1.4.5),
- over-regional aspect (top-down approach),
- cross-sectorial (over-sectorial) aspect.

Landscape-ecological Complexity—Integrated Approach

The landscape-ecological complexity—the integrated approach to the landscape—requires that the aforementioned confrontation of societal activities is performed in the logical as well as a temporal sequence with respect to the following:

- the properties of the primary landscape structure (particularly abiotic landscape elements)—these conditions are physically most **constant** and their basic principles are more or less unchangeable;
- the properties of the secondary (current) landscape structure (especially the current flora and fauna, anthropogenic structures, land use)—these conditions are also physically fixed at one place and their **change requires expending energy**;

- the properties of the tertiary structure (particularly socio-economic factors/phenomena and processes in the landscape)—these are not or only slightly physically bound to a specific place and their change is not induced by physical energy but rather **decisions**. They include plans, concepts, agreements, conventions, law, and economics—physically intangible elements.

From the perspective of ecological optimisation, the most substantial are the unchangeable and partially changeable landscape-ecological conditions—primary and secondary landscape structure—whose disturbance is the crux of all problems, while the tertiary landscape structure and new requirements must be adapted to the first two landscape structures. The extent of the analysis of the properties of the landscape structures strictly depends on the hierarchical level of a specific task.

Over-Regional Aspect

The over-regional aspects mean that landscape-ecological conditions should be assessed with respect to systematically higher hierarchical structure: over-regional–regional–local spatial structures. It is a top-down principle of spatial sequence and territorial differentiation of interests, i.e. the harmonisation of interests from the level of state through general plans, large territorial units, regions, municipalities, up to functional zones (as, e.g. residential, production, recreational and other zones). In other words, the lower levels of planning documents must adopt the principles and regulations established at the higher levels and not the other way round, to prevent disorganisation of spatial arrangement as well as land use.

This sequencing complies with the **hierarchical levels of territorial planning,** which is legally obliged to employ the LANDEP outputs.

As territorial units gradually become smaller, methodological procedures are more elaborated and detailed.

Cross-Sectorial (Over-Sectorial) Aspect

The cross-sectorial (over-sectorial) aspect specifies that the optimisation process includes a confrontation the properties of a landscape system with the **entire complex of demanded social activities**.

In one sentence: The LANDEP includes the confrontation and harmonisation of:

- **all properties of a landscape**, with the
- **entire complex of demanded social activities** for the
- **entire concerned territory**.

This represents a significant principle of a decision-making process in LANDEP. LANDEP should not a priori impede any activity demanded by society, but rather seek harmony between ecological conditions and a particular activity.

This principle can be referred to as a **principle of a least evil,** i.e., we try to find an area for each activity where it will be in minimum collision with the natural conditions.

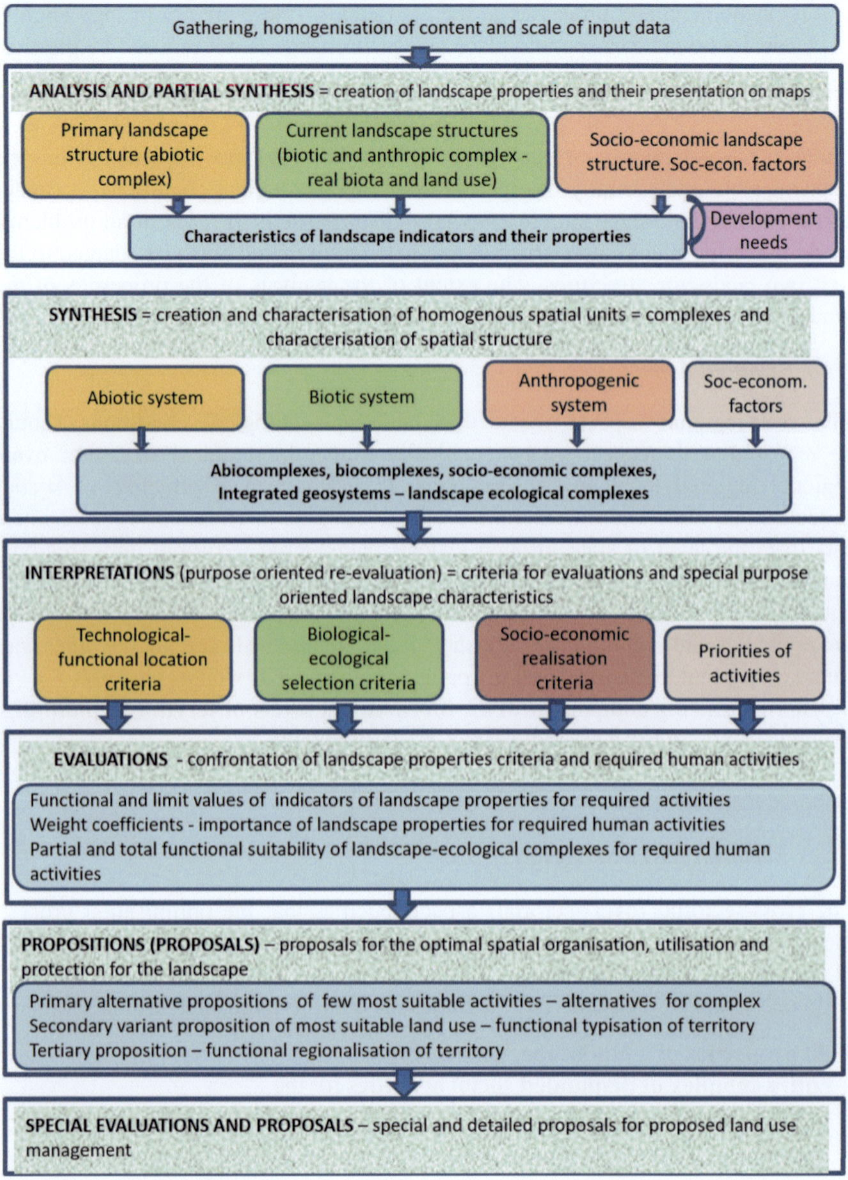

Fig. 2.3 The LANDEP scheme

2.3 The Process of the LANDEP

Since its first adoption (Ružička and Miklós 1979), the LANDEP method has experienced many improvements. As different experts have been involved in the process of its development, partial methods and procedures have notably been improved within individual steps, especially in the process of interpretation and evaluation. Nevertheless, it is yet important to mention that the methodological basis, subject and chronological sequence of the steps, the logic of confrontation between the landscape and man's requirements, and the direction of the output have remained virtually unchanged.

The methodology and methods in LANDEP can theoretically be derived into elementary steps from reverse analysis from final goals to start as follows:

- The ultimate goal of LANDEP is the ecologically optimal functional organisation of the landscape in the form of a proposal for the optimal localization of social activities in the landscape—the step marked as **propositions**.
- To be able to propose an optimal organisation and use of the landscape, it is necessary to determine the suitability of each element of the landscape for different activities from different perspectives—the step marked as **evaluation**.
- For the purpose of evaluation, we need some evaluation-relevant—not just whatever—information about the landscape. To this end, a step for development of special-purpose criteria—interpreted indicators of landscape features—has been developed. The interpreted indicators are derived from the basic analytical indicators or the indicators of synthetic spatial units. This step is called **interpretation**.
- The interpretations are based on various combinations of analytical features within the same element or area of the landscape. Therefore, we need to describe the landscape features comprehensively—in complex homogeneous spatial units—creating types and regions of landscape-ecological complexes, within the process called a landscape-ecological **synthesis**.
- For the purpose of the synthesis, we must obtain the necessary information about the landscape properties. The process of their acquisition, modification and homogenization are represented by the step named a landscape-ecological **analysis**.

Obviously, the real landscape-ecological planning proceeds from analyses to propositions, as shown in Fig. 2.3.

This is how we describe the steps of the LANDEP in the following chapters.

References

Deverre Ch, De Sainte Marie Ch (2008) L'écologisation de la politique agricole européenne. Verdissement ou refondation des systèmes agro-alimentaires? Revue d'Etudes en Agriculture et Environment 89(4):83–104

Miklós L (1991) Zásady slovenskej ekologickej politiky. Životné prostredie 25(4):174–178 Bratislava

Miklós L (2011a) The environmental policy of the Slovak Republic after the Velvet Revolution. Životné prostredie 45(6):295–300

Miklós L (2011b) The landscape in the international and Slovak environmental politics. Životné prostredie 45(6):307–312

Ružicka M, Miklós L (1979) Teoretické a metodologické základy biologického plánovania krajiny. Záverecná správa úlohy VI-3-5/1. Bratislava, ÚEBE SAV

Ružička M, Miklós L (1982a) Metodické poznatky ekologického hodnotenia územia pre zónu a sídelný útvar (na príklade Rimavskej Soboty). Acta Ecologica 9(26):74 (VEDA, Bratislava)

Ružička M, Miklós L (1982b) Landscape-ecological planning (LANDEP) in the process of territorial planning. Ekológia (ČSSR), l:297–312

Ružička M, Miklós L (1990) Basic premises and methods in landscape-ecological planning and optimisation. In: Zonnenveld IS, Forman RTT (eds) Changing landscapes: an ecological perspectives. Springer, New York, pp 233–260

Chapter 3
Landscape-ecological Analyses in LANDEP

Abstract In the planning processes, there is usually no time devoted for own analysis, therefore, in the planning existing analytical materials are used that should be adapted to their particular needs. **Re-assessment and homogenization** of the existing results of a wide variety of specialised research are the objective and the content of the LANDEP analyses. The analyses in LANDEP are arranged according to the structures of the landscape as a geosystem. However, the very first input into the LANDEP process is defined by delimitation of the area of the territory of interest. Analyses of the **primary landscape structure** relate, in particular, to the analysis of the abiotic elements of geosystems. The most frequently used indicators of landscape properties for deciding on the optimal landscape use are to be chosen from the following subcomplexes:

- subcomplex substrate–soil–underground water: most commonly used indicators are depth, skeletality and texture of the substrate and soils, subsurface water depth;
- subcomplex relief—slope dynamics: especially slope inclination, curvature and relief forms, topographic position, slope length;
- subcomplex relief—dissection: morphological and morphometric relief types.
- subcomplex surface water: hydrographic characteristics of river-basins and their subdivision, hydrological characteristics of flows and lakes;
- subcomplex (micro) climate: especially relief orientation to the cardinal points, insulation and shadowing, characteristics of climate of geographical regions.

The secondary landscape structure is for the purposes of LANDEP referred to as **the current landscape structure**. It consists of the elements that are partially or fully affected, modified or made by man and currently cover the earth surface These elements are characterized most often according to their

- functional character as land-use forms,
- biological–ecological content, as physiognomic-ecological formations of actual vegetation,
- biotopes, or habitats for fauna.

L. Miklós and A. Špinerová, *Landscape-ecological Planning LANDEP*,
https://doi.org/10.1007/978-3-319-94021-2_3

Analyses of the **tertiary landscape structure** deal with the socio-economic factors/phenomena, which are nonmaterial, intangible in nature, they only occur in various laws, standards, regulations and other development documents. Nevertheless, they can be mapped or projected onto a map, and thus allow the determination of their spatial encounters and even conflicts in the landscape. According to their nature, SEF mapping units have the form of the following:

- borders and territories of nature conservation and protected natural resources areas;
- protective, safety and sanitary zones of production, transport and other technical facilities;
- borders of administrative units and borders of other sectorial territorial units;
- borders of areas of declared territories with deteriorated state of the environment.

A specific part of the analyses is devoted to the definition of the desired human activities which enter to the planning procedure. The chapter is illustrated by numerous tables and maps.

Keywords Primary · Secondary · Tertiary landscape structure · Activities

The landscape-ecological analyses are the input into all following step of LANDEP up to the decision-making process on the optimal use of the landscape. Concretely, they primarily serve as **sources** necessary for syntheses.

In the planning processes there is usually no time devoted for own analysis, therefore in the planning are used existing analytical materials that should be adapted to their particular needs. **Re-assessment and homogenization of the existing** results of a wide variety of specialised research are the objective and the content of the LANDEP analyses.

However, analyses are not available, or not in requested quality; therefore, they may be conducted for a particular task individually. A crucial precondition for a successful process of a complex landscape ecological planning process is a high-quality geographical information system (Verrasztó 1979, 2010; Burrough and McDonnel 1998; Bandrova et al. 2010). Moreover, a proper spatial information system must be based on the understanding of the landscape as a geosystem (Ružička and Miklós 1982, 1990; Miklós and Izakovičová 1997; Miklós and Špinerová 2011).

The objective and content of the analyses is the
re-assessment and homogenization
of the existing results of a wide variety of specialised research.

An imperative and inevitable starting point is to differentiate the basic terms used to mark the different kinds of information. Accordingly, it is to distinguish the following sequence of terms and their content.

Geosystem consists of **elements** that we examine through the **indicators** of their **properties** that have specific **values** in each point of the landscape (Csorba and Szabó 2012). A strict conceptual distinction between the terms in the chain '***geosystem–elements–property–indicator–values***' in each point of the space is absolutely crucial. This approach then dictates that the objective of the LANDEP analyses is to create and characterize property indicators of the landscape elements. Indicators are described as follows:

- state variables,
- typological characteristics of spatial units—spatial subsystems,
- a combination of typological characteristics and state variables.

In the publication, the indicators of the properties of the elements of landscape (geosystems, complexes) are marked as x_n, y_n, z_n for easier illustration of the relations in the LANDEP process.

The LANDEP analyses include the following steps.

3.1 Delimitation of the Area of the Territory of Interest

This is the very first input into the LANDEP process and is defined by delimitation of the area of the territory of interest. It is appropriate to delimit the area as a whole, with natural borders that try to follow and cover also the borders of administrative or other sectorial territorial units. The larger and more diverse the area is, the more successful the decisions become as it provides greater room for optimal distribution of activities. This means that the ecological optimization of the organisation and utilisation of the landscape should proceed top down.

It is, thus, appropriate to prepare a cartographic projection base with the main topographical features. These cartographic projection bases should be used as the basis for spatial information system for the projection of all materials. The projection of all the processed information into a single groundwork is a prerequisite for correct syntheses, as well as for the further steps of the methodology to be successful (Kocsis et al. 2016b).

An illustration of a traditional unified cartographic base for LANDEP planning is in Fig. 3.1. The cartographic base served as a unified hand-made GIS reference surface, where all other spatial information were projected (Miklós et al. 1986).

Currently, the GIS technologies and the remote sensing data are widely used also for creation of cartographic base (Fig. 3.2, Špinerová 2011).

According to objects and methods, the analyses are pragmatically divided into the following.

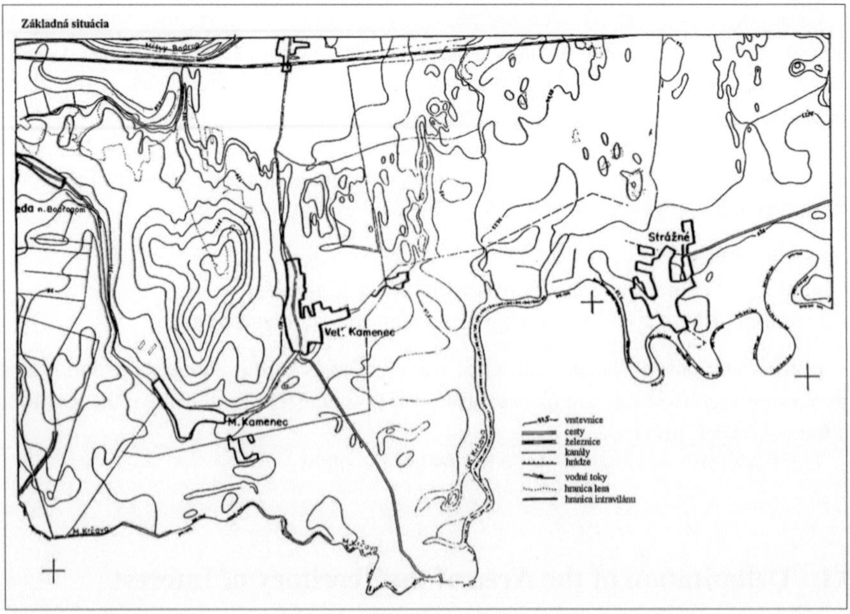

Fig. 3.1 An illustration of a traditional unified cartographic base: basic topographic elements redrawn by hand into a matrix—Eastern Slovak Lowland—section

3.2 Analyses of the Primary Landscape Structure

This part of analyses relate in particular to the analysis of the abiotic elements of geosystems.

The existing experience has shown that the most frequently used indicators of landscape properties for deciding on the optimal landscape use are to be chosen from the following subcomplexes of the primary landscape structure (Bedrna et al. 1992; Bedrna 2002):

- subcomplex substrate–soil–underground water

Most commonly used indicators are: depth, skeletality and texture of the substrate and soils, subsurface water depth;

- subcomplex relief—slope dynamics: especially slope inclination, curvature and relief forms, topographic position, slope length;
- subcomplex relief—dissection: morphological and morphometric relief types.
- subcomplex relief—surface water: hydrographic characteristics of river-basins and their subdivision, hydrological characteristics of flows and lakes
- subcomplex relief—(micro)climate: especially relief orientation to the cardinal points, insulation and shadowing, characteristics of climate-geographical regions.

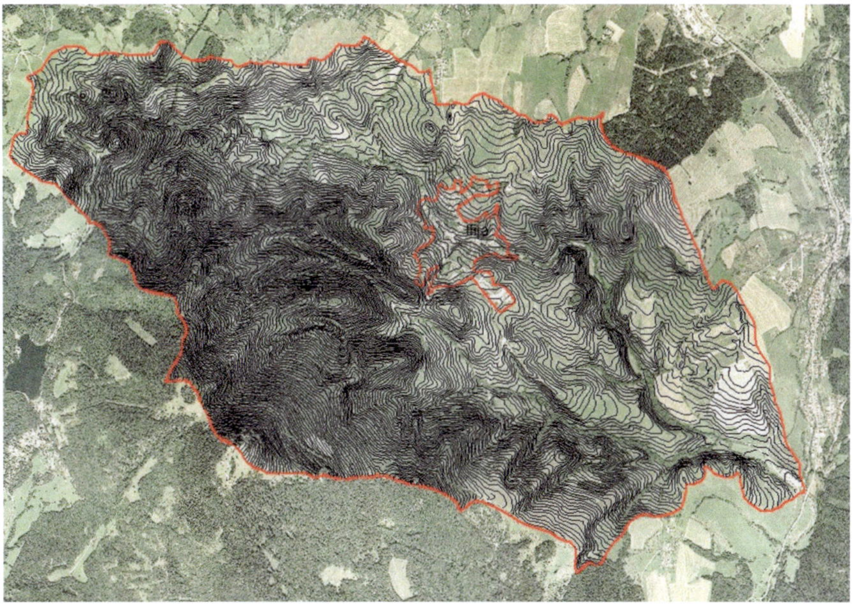

Fig. 3.2 An illustration of current cartographic base: a combination of orthoimages and topographic base with GIS technology—Ilijský stream catchment

Examples of often used classification classes are shown in Tables 3.1, 3.2, 3.3 and 3.4.

The following pages are illustrated by examples of analytical maps from a model territory of Ilijský creak catchment area (Figs. 3.3, 3.4, 3.5, 3.6, 3.7 and 3.8) (Špinerová 2005, 2010).

Note: All maps shown in the next chapters are processed electronically in GIS techniques on resolution level of mapping scale 1:10,000. The maps on figures in this publication are considerably reduced by size and content, they serve just as illustrations of the described process. The valid explanation of the methodical content of the figures related to this publication is presented under the title of figures.

3.3 Analyses of the Secondary Landscape Structure

As we mentioned in the theoretical part (Sect. 3.3), the secondary landscape structure, as a result of land use, is for the purposes of LANDEP referred to as **the current landscape structure** (**CLS**). CLS consists of the elements that are partially or fully affected, modified or made by man and currently cover the earth surface (Miklós and Izakovičová 1997). These elements can be characterized, at different levels of detail most often according to their

Table 3.1 Typological mapping units of the geological–substrate complex

	Genetic group of sediments	Type of geological–substrate complex
01	Fluvial sediments	Fluvial sediments with organic addition
02		Clay-loam to clay floodplain sediments
03		Loam to sandy-loam floodplain sediments
04		Sandy to loam-sand floodplain sediments
05		Sand-loam terrace sediments
06		Gravel-loam to gravel floodplain sediments
07		Gravel-loam terrace sediments
08	Proluvial sediments	Sand-loam to sandy proluvial sediments
09		Gravel-loam to gravel proluvial sediments
10	Aeolic sediments	Loess to loess-loam
11		Drifting sands
12	Deluvial sediments	Clay to sand-loam deluvial sediments
13		Gravel-loam deluvial sediments
14	Glacifluvial sediments	Sand-gravel glacifluvial sediments
15		Gravel-boulder glacifluvial to glacial sediments
16		Boulder glacial sediments (moraine)
17	Other sediments	Gravel-boulder colluvial sediments
18		Organogenic sediments (peat)
19		Travertine
20		Landslide
21		Anthropogenic sediments
22	Local weatherings, eluviums and deluviums	Silt-loam to clay weathering in non-hardened clay sediments (tertiary clay and silt clay)
23		Gravel-sand in non-hardened gravel sediments (tertiary gravels with deposits of sands)
24		Loam weatherings in hardened clay-silt rocks (shales, clay, limestone and marlstone)
25	Loam to stone weatherings on different rocks	Flyschoid rocks (shales, sandstone, limestone and marlstone)
26		Sandstone-conglomerate rocks (sandstone, conglomerate, shales, quartzite, arkoses, limestone and marlstone)
27		Massive limestone-dolomite rocks (limestone, dolomite)
28		Volcanic pyroclastics (tuff, tuffs)
29		Effusive rocks (melaphyre, rhyolite, rhyodacites andesite, basalite)
30		Intrusive magmatic rocks (granites, granodiorite, diorite)
31		Low metamorphic rocks (phyllite, phyllonite)
32		Highly metamorphised rocks (gneiss, mica schist, amphiboles, migmatites)

Table 3.2 Classification classes of the selected state variables of soils

Categories of soil texture groups—soil grain size		
	Grain size	Soil texture group name
01	Sand	Light soil
02	Loam–sand	Light soil
03	Sand–loam	Moderately heavy soil
04	Loam	Moderately heavy soil
05	Silt–loam	Moderately heavy soil
06	Silt	Moderately heavy soil
07	Sand–clay–loam,	Moderately heavy soil
08	Clay–loam	Moderately heavy soil
09	Silt–clay–loam	Moderately heavy soil
10	Sand–clay	Heavy soil
12	Clay	Heavy soil
Categories of soil depth		
	Soil depth (cm)	Depth category of soil
1	Over 61	Deep
2	31–60	Moderately deep
3	Up to 30	Shallow
Categories of soil skeletality		
	Skeletality of soil (%)	Skeletality category of soil
1	Up to 25	Without or scarcely stony
2	25–50	Moderately stony
3	over 50	Very stony

- functional character as land-use forms, to their
- biological–ecological content, as physiognomic-ecological formations of actual vegetation, as well as
- biotopes, or habitats for fauna.

They have the nature of topical spatial units, and from this perspective, they are often marked as **CLS mapping units**.

A basic classification of mapping units of CLS still follows the original division of the secondary landscape structure, as proposed by Ružičková and Ružička (1973). The differences are only in the level of detail of the work that is based on its objectives and the overall character of the current landscape structure of the model territory.

Examples of mapping units at different levels are shown in Tables 3.5, 3.6 and on Fig. 3.8.

Table 3.3 Typological characteristic of soils—the most frequently used soil associations in LANDEP

	Soil association
1.	Leptosol
2.	Regosol
3.	Rendzic Leptosol (Rendzina)
4.	(Para)rendzic Leptosol (Pararendzina)
5.	Podzol
6.	Cambisol
7.	Pseudogley (Stagnosol)
8.	Albic Luvisol (illimerized soil)
9.	Vertisol
10.	Luvisol
11.	Chernozem
12.	Mollic Fluvisol (arenic)
13.	Fluvisol
14.	Gleysol
15.	Organosos
16.	Solonetz
17.	Anthrosol
18.	Technosol

3.4 Analyses of the Tertiary Landscape Structure

As mentioned in the theoretical–methodological part of the work (Sect. 1.4.4), the socio-economic phenomena, SEF, are non-material, intangible in nature, they only occur in various laws, standards, regulations and other development documents. Nevertheless, they can be mapped or projected onto a map, and thus allow the determination of their spatial encounters and even conflicts in the landscape. The SEF from the ecological–environmental point of view might display the landscape areas with positive measures, especially SEF of natural resources protection and nature conservation, as well as areas where negative impact occurs, characterized as stress factors.

According to their nature, SEF mapping units have the form of the following:

- borders and territories of declared nature conservation areas and protected nature resources areas;
- protective (buffer) zones, safety zones and sanitary zones of production, transport and other technical facilities;
- borders of administrative units and borders of other sectorial territorial units;
- borders of areas or sections of declared territories with deteriorated state of the environment.

Table 3.4 Classification classes of the morphological-morphometric relief types

Morphological-morphographic types—characteristics		Position
01	Flat top	Initial
02	Plateau top	Initial
03	Dome-shaped top	Initial
04	Ridge	Initial
05	Compact dell	Transport
06	Ground of dell valley	Transport
07	Ground of slope valley	Transport
08	Compact slope valley	Transport
09	Transport slope	Transport
10	Slope platform	Transport
11	Large accumulation plain	Transport
12	Wavy plain of sand dune	transport
13	Significantly wavy slope	transport
14	Slope divided by gullies and small valleys	Transport
15	Terrace (river)	Transport
16	Narrow floodplain of mountain streams	Transport/Final
17	River channel in section (natural and anthropogenic)	Transport/Final
18	More significant elevation as part of floodplain	Final
19	Slope in foothill position/foothill	Final
20	Alluvial cone	Final
21	Wide riverside floodplain	Final
22	Closed depression en bloc (a-closed, b-oxbow)	Final
23	Bottom of waterlogged depression	Final
Specific forms of relief		
24	Peak	Initial/Temporary
25	Mountain saddle	Temporary
26	Sliding slope	Transport/Temporary
27	Terraced slope	Transport/Temporary
28	Scree and deposit of stones (stone sea)	Transport/Temporary
29	Rock scarp	Transport/Temporary
30	Bottom of glacial trough	Transport/Temporary
31	gully to ravine	Transport/Temporary
32	Moraine mound	Transport/Temporary
33	Glacial cirque	Final/Temporary
34	Bottom of water reservoir	Final/Temporary
35	Anthropogenic form	Undefined/Temporary

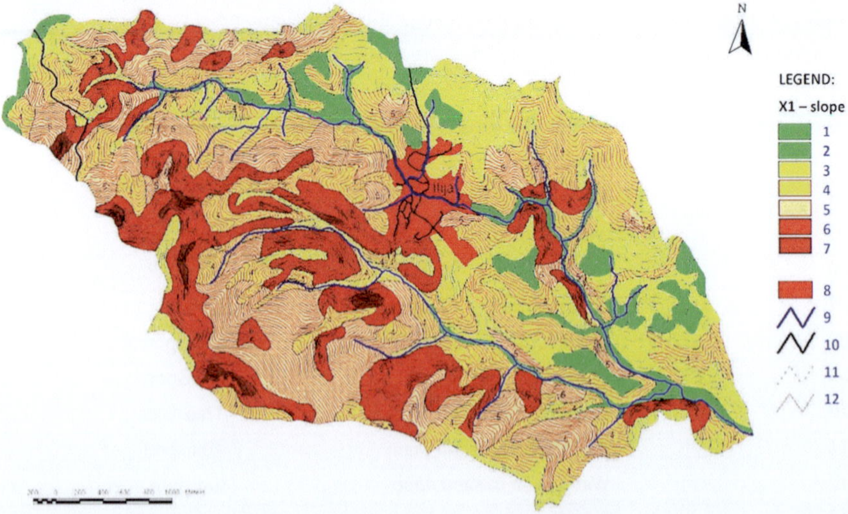

Fig. 3.3 Illustrations of analytical maps: indicator of a biocomplex ABC x_1—slope angle. Legend: x_1—slope angle: 1—⟨0°–1°⟩; 2—⟨1°–3°⟩; 3—⟨3°–7°⟩; 4—⟨7°–12°⟩; 5—⟨12°–17°⟩; 6—⟨17°–25°⟩; 7—⟨more than 25°⟩; Topography: 8—intravilan, 9—watercourse, 10—asphalt road, 11—unpaved path, 12—contour line

A large part of the mapped SEF is related to the man-made material elements of the current landscape structure CLS. Some elements of the CLS are regarded as primary stress factors, they are bound mostly with SEF of the character of protective zones or safety zones (Šúriová and Izakovičová 1995; Izakovičová et al. 2000) (see Table 5.13 in Sect. 5.3).

Other SEF are of very different characters. They are bound to

- nature conservation (Table 3.7)
- protection of natural resources (Table 3.8)
- urban, residential and recreation areas (Table 3.9)
- communication and transport of industrial and mining activities, agricultural activities (Table 3.9)
- deterioration of natural resources and the environment (Table 3.10)
- territorial units—administrative boundaries (Table 3.11).

The tables show the most frequent SEF from above-listed groups, in particular, those which are declared by law and other planning and development documents. They are listed in the tables in relation to the material elements and structures of the primary and secondary structure of landscape.

Summary of the Analyses of the Landscape Structures

It is obvious that different tasks may request analyses of different indicators. Analyses vary with areas and tasks. Important criteria for the analyses are the following:

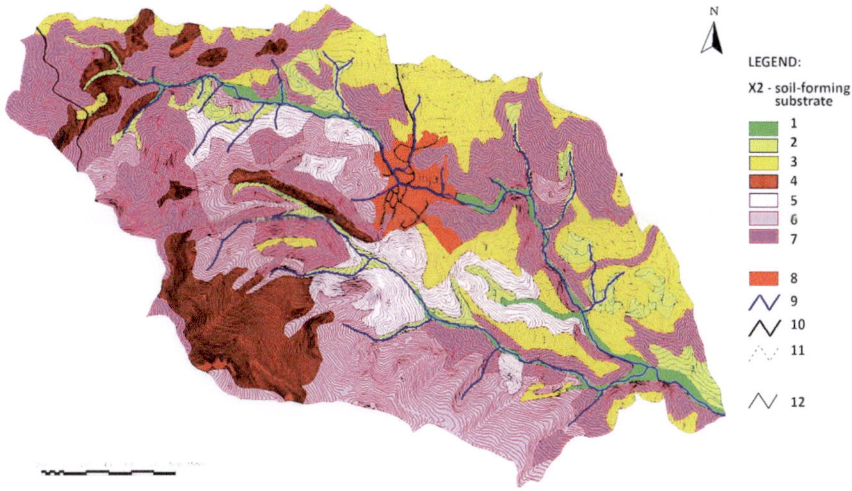

Fig. 3.4 Illustrations of analytical maps: indicator of a biocomplex ABC $\boldsymbol{x}_2$—soil forming substrate. Legend: Fluvial sediments: 1—fluvial sediments (gravel-loam to rocky); Deluvial and subslope sediments (eluviums, deluviums, colluviums): 2—sediments of dells and valleys bottoms; 3—slope platforms deluvial sediments, slope loams; 4—sedimentary debris (rocks, boulders); Weatherings: 5—on volcanic sandstones, silts and pumices; 6—on breccias; 7—on andesites, porphyries and rhyodacites. Topography: 8—intravilan, 9—watercourse, 10—asphalt road, 11—unpaved path, 12—contour line

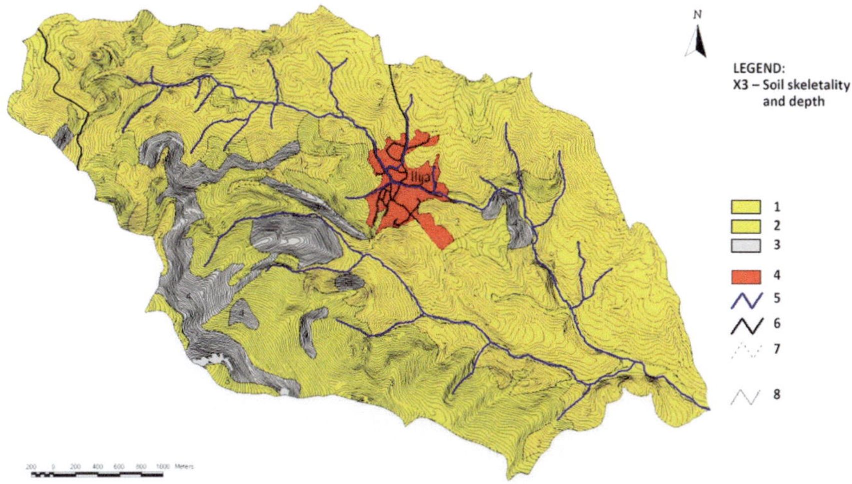

Fig. 3.5 Illustrations of analytical maps: indicator of a biocomplex ABC $\boldsymbol{x}_3$—soil skeletality and depth. Legend: 1—weakly to moderately skeletal, moderately deep to deep; 2—moderately skeletal, moderately deep; 3—moderately to very skeletal, shallow; Topography: 8—intravilan, 9—watercourse, 10—asphalt road, 11—unpaved path, 12—contour line

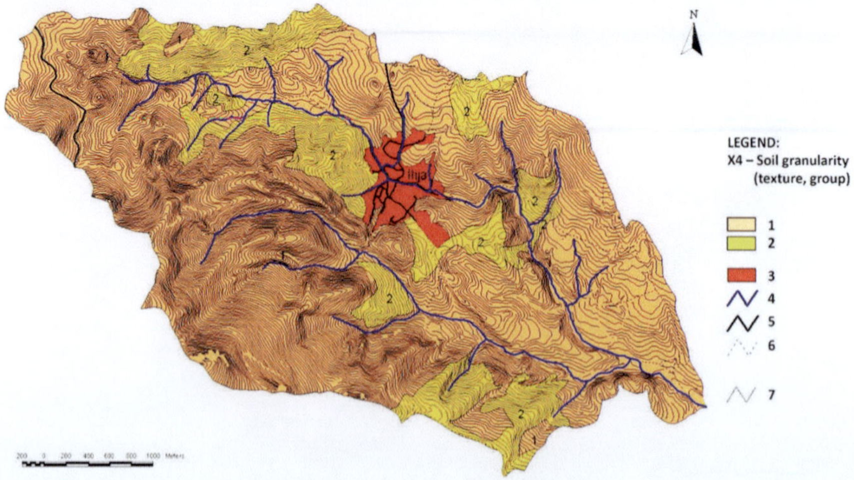

Fig. 3.6 Illustrations of analytical maps: indicator of a biocomplex ABC x_4—soil texture (grain size). Legend: 1—loam; 2—sandy-loam; Topography: 8—intravilan, 9—watercourse, 10—asphalt road, 11—unpaved path, 12—contour line

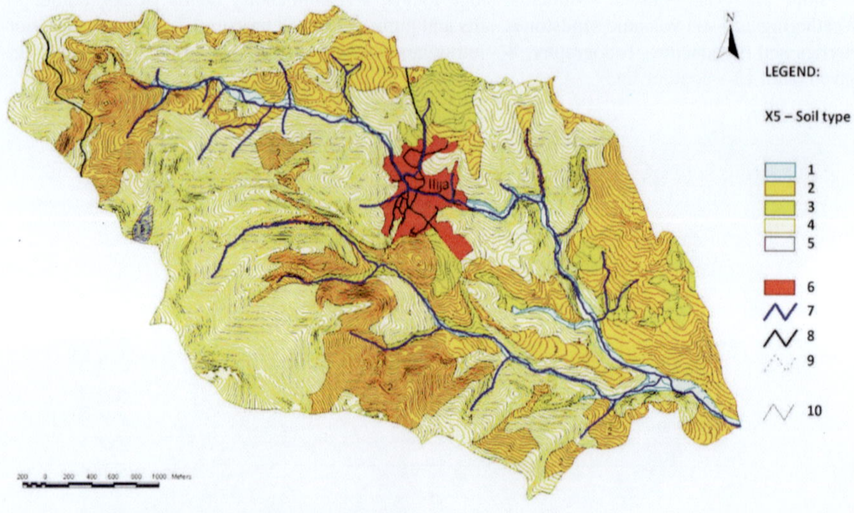

Fig. 3.7 Illustrations of analytical maps: indicator of a biocomplex ABC x_5—soil types and subtypes. Legend: 1—Haplic Fluvisol (typical), 2—Haplic Cambisol (typical), Leptic Luvisol, 3—Stagnic Cambisol, Fragic Cutanic Albeluvisol, 4—Haplic Cambisol (typical), 5—Fulvic Andosol (typical). Topography: 8—intravilan, 9—watercourse, 10—asphalt road, 11—unpaved path, 12—contour line

- indicators should explain the he diagnosis of the properties, they can be parametrized and coded;
- they can be mapped throughout the study area;

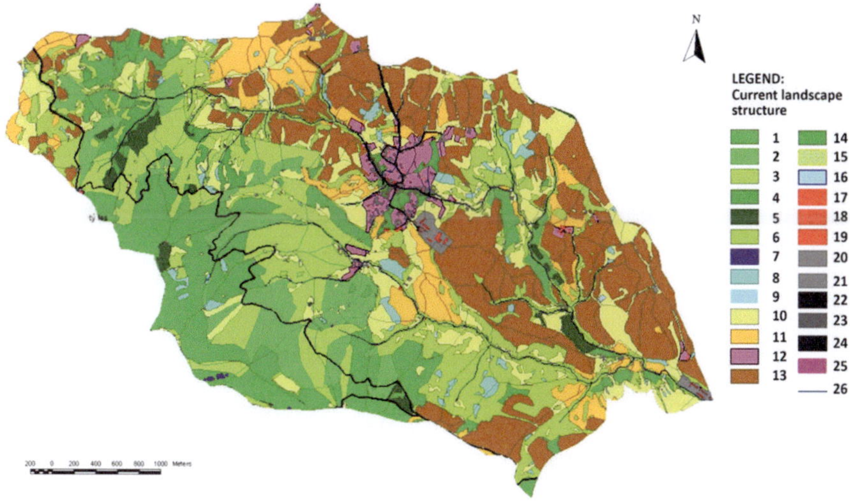

Fig. 3.8 Illustrations of analytical maps: elements of the secondary landscape structure CLS y_1—land use and land cover. Legend: 1—broadleaf forest; 2—mixed broadleaf forest; 3—mixed forest; 4—mixed coniferous forest; 5—coniferous forest; 6—unknown forest; 7—rock; 8—group of shrubs; 9—water reservoir; 10—grassy and scrubby vegetation; 11—meadow, pasture; 12—orchard; 13—arable land; 14—settlement vegetation; 15—playground; 16—swimming pool; 17—separate building; 18—tower; 19—hut, shed; 20—non-specified area of the buildings; 21—material and waste landfill; 22—road; 23—parking lot; 24—water tank; 25—terrain slit; 26—watercourse

- they differentiate the study area, i.e. they have various values within the territory. On the contrary, those indicators that have the same value throughout the territory, i.e. they do not segment the area, are of no importance.

The results of analyses are usually

projected onto maps

with coded values of indicators of landscape properties, which serve as the basis for synthesis. Often formalised values and various landscape metrics are applied, too (Džatko et al. 1976; Hraško et al. 1991; Herzog and Lausch 1999; Szabó and Csorba 2009; Konečný et al. 2012).

3.5 Desired Social Activities ‘R’

The economic sector and other spheres of social activities demand certain amount of area in the landscape for their development, either existing or new areas. The demands of the sectors are realised first by occupation of certain area, than by change

Table 3.5 Mapping units of the current landscape structure CLS at different levels

Mapping units on different level		
1:500,000	1:200,000, 1:100,000	1:50,000, 1:25,000
Forests and greenery	Broad leaved forests	Flooded forests
		Oak forests
		Beech forests
		Linden-maple forests
		Other broad leaved forests
	Coniferous forests	Dwarf mountain pine
		Natural spruce forests
		Natural fir forests
		Natural pine forests
	Mixed and other forests	Mixed forests (by composition)
		Forest nurseries
	Monocultures	Monocultures (by composition)
	Landscape greenery	Tree lines and solitaires
		Scrub vegetation
		Wetlands, moorlands, peat lands
Types of agriculture	Permanent grassland	Permanent grassland of different character (wet, dry, submountain, etc.)
		Ruderal vegetation
	Arable land	Arable land with small blocks
		Arable land with big blocks
	Permanent crops/cultures	Gardens, orchards, plantations
		Vineyards
		Hop gardens
		Greenhouses, foil fields
Waters	Water areas and water flows	Lakes, glacial lakes
		Water reservoirs
		Natural water flows and channels
		Regulated water flows and channels
Exposed substrate	Exposed substrate	Rocks
		Slope debris, rock slopes, block fields
		Gravel pits, sandpits
Urban areas	Recreational areas	Recreational areas
		Recreational objects
		Sports areas, recreational fields

(continued)

Table 3.5 (continued)

Mapping units on different level		
1:500,000	1:200,000, 1:100,000	1:50,000, 1:25,000
	Residential areas	Parks and public greenery
		Service areas
		Residential areas
Technical objects	Industrial areas	Mining objects and areas
		Object and areas of energetical industries, incinerators
		Objects and areas of production
		Warehouses
		Other
	Objects and lines of infrastructure	Traffic lines and objects
		Objects of logistic
		Telecommunication and power transmission lines
		Objects and lines of pipelines
	Landfills and heaps	Ponds
		Heaps
		Landfills
	Military objects	Military objects and areas
	Agricultural objects	Agricultural–technical objects
		Stables
		Sheepfolds
		Bioenergetical objects, biomass processing objects
		Dumps of fertilizers

of its utilisation, its structure, by building new objects, and by defining function zones that then serve as places for performing the given activity. Areas are, thus, required for **conducting a particular activity** needed in a particular sector. For landscape-ecological planning (but also for other types of planning), it is necessary to project these activities onto maps, plans and projects. The activities are, of course, **dynamic**, the maps are by their character **static**, thus to capture the dynamic character of activities to maps is impossible. Therefore, the requested 'activities' on maps are projected statically as **areas of activities**. After their realisation these areas of activities become the **elements of the current landscape structure CLS**, in the form of functional elements of land use, elements of actual vegetation, habitats, etc. The existing areas with a certain activity are, of course, referred to as the elements of current landscape structure, too.

Table 3.6 Mapping units of current landscape structure CLS on the local level: example of Ilijský creak catchment area

i	Unit of CLS	Area in m^2	% of the total area
1.	Broadleaf deciduous forest	2,260,875	14.5
2.	Mixed broad-leaved deciduous forest	1,010,146	25.4
3.	Mixed forest with balanced stand composition	3,966,003	6.5
4.	Mixed forest with prevalence of coniferous trees	215,643	1.4
5.	Coniferous forest	244,187	1.6
6.	Not definable mixed forests	300	0.00
7.	Rocks, rock slopes, terrain notch	1738	0.04
8.	Continuous shrub vegetation	302,152	1.9
9.	Water reservoir	500	0.00
10.	Mosaic of grasslands and scrub vegetation	2,101,958	13.5
11.	Meadows and pastures	1,091,530	7.0
12.	Orchards	334,157	2.1
13.	Arable land	3,646,209	23.4
14.	Greenery in residential areas	97,180	0.6
15.	Sports areas	1236	0.01
16.	Other technical elements	52,384	0.3
17.	Separate buildings	500	0.00
18.	Other built-up areas in settlements, courtyards	157,479	1.0
19.	Non-specified areas in settlements	500	0.00
20.	Landfills	4785	0.03
21.	Reinforced structures (roads, parking places)	109,853	0.7
22.	Terrain notch	200	0.00
23.	Water courses	3540	0.02

Into the decision-making process enters the abstract, at that time only required, planned activities, which do not yet have any area for being performed, and only require some. However, after their projection on a map, they take form of an area of activity, thus they become a concrete element of the proposed CLS. This sequence can be expressed as follows:

Table 3.7 Frequently used typological characteristics of the SEF bound to the nature conservation declared in acts and other planning and development documents

Elements of primary and secondary landscape structure	SEF defined in various regulations
Specific landscape structure of a predominantly natural character	Different categories of national network of nature conservation areas and their buffer zones • Protected Landscape Area (PLA) • National Park (NP), NP protection zone (PZ NP) • Protected Area (PA), PA protection zone (PZ PA) • Nature Reservation (NR) • National Nature Reservation (NNR), buffer zone NR and NNR • National Natural Monument (NNM) • Nature Monument (NM), caves, natural waterfall • Protection zone of NM and NNM • Private protected areas
	Zones of PLA and NP
	NATURA 2000 • Special Areas of Conservation—Sites of Community Importance (SCI) • Special Protection Areas—Protected Bird Areas (PBA)
	Sites of the UNESCO World Cultural and Natural Heritage
Significant landscape structures and elements	Biospheric reserves of the UNESCO's Man and Biosphere Programme (MAB UNESCO)
	Wetlands by Ramsar convention
	Elements of TSES in development documents, plans and projects • Protected landscape element—biocentre TSES • Protected landscape element—biocorridor TSES • Protected landscape element—interactive TSES element
	Historical landscape structures
Important species and trees	Gene sites of flora and fauna Protected plants, animals Protected trees (PT) and their buffer zones Protected minerals, fossils/petrifications

Table 3.8 Frequently used typological characteristics of the SEF bound to the protection of natural resources declared in acts and other planning and development documents

Elements of primary and secondary landscape structure	SEF defined in various regulations
Forests	Protective forests • In exceptionally unfavourable places • High mountain forests, forests in the scrub pine belt • Other forests with predominant protection of soil
	Forests of specific designation • In protective zones of water resources • In protective zones of healing and mineral resources (spa forests) • In nature conservation areas • Forests designated for research • Forests with medical-recreational function, nearby medical institutions • Suburban forests with health-recreational function, forest parks • In recognized animal and pheasant grounds • Forests under the influence of emissions
	Commercial and other forests • Timber production forestry • Forest nursery • Military forests • Forests land in Sites of Community Importance • Forest land in declared Protected Bird Area
Water	Protected water area (protected area of natural accumulation of water)
	Water-supply rivers and reservoirs
	Catchment of water-supply rivers and reservoirs
	Important water flows
	Zones of hygiene protection of underground and surface water of 1st to 3rd degrees (protective zone of water sources)
	Protective zones of reservoirs
	Water suitable for various functions: bathing, irrigation, for fish breeding and reproduction of original species of fish
Waters with specific properties (natural healing sources)	Protective zones 1–3 levels for natural healing water sources
	Natural healing sources of gases and emanations
	Natural healing resources from peat fens, sludges and other earth healing sources

(continued)

Table 3.8 (continued)

Elements of primary and secondary landscape structure	SEF defined in various regulations
	Protective zones 1–3 levels for natural sources of mineral water
	Spa cities (natural healing spa), protective zone of spa cities
	Interior spa territory
	Exterior spa territory
Climatic conditions beneficial for healing	Protective zone of climatic condition beneficial for healing
Agricultural soils	Registered parcel, culture, notes on culture
	Bonited soil-ecologic units (BSEU)
	Registered permaculture: vineyard, orchards, permanent grassland for grazing, growing areas
	Agricultural production areas
	Investment to land (melioration, irrigation)
	Agricultural land in Sites of Community Importance
	Less-favoured areas in registered soil blocks
Habitats for fish and animals	Protected fishing areas
	Fishing grounds
	Fish-farming flows
	Recognized animal keepers
	Pheasantries
Minerals	Protected area of deposits with reserved minerals
	Protected area of deposit of non-reserved minerals
	Natural rock structures for storing gas
Historic structures and sites	Cultural monuments and national cultural monuments
	Memorial City Reservation (MCR) and Folk Architecture Reservation (FAR)
	Monuments zone—city and landscapes
	Archaeological find
	Protective zones of the memorial fund

Table 3.9 Frequently used typological characteristics of the SEF bound to the zones of urbanisation, sanitary and safety zones of industrial, transport, technical and communal activities declared in laws and other planning and development documents

Elements of primary and secondary landscape structures	SEF defined in various regulations
Residential areas	Residential zones of individual and complex apartment building
	Zones of public services
	Cultural and school zones
	Zones of relaxation and sport, parks and public greenery
	Protective zones of hospital areas
Recreational areas	Declared recreational area, zones and centres of varying levels
	Declared campgrounds
Production areas	Hygienic protective zones (HPZ)/Safety zones (SZ) of industrial areas areas
	HPZ/SZ of quarries and mining sites
	HPZ/SZ of waste landfills
	HPZ/SZ of wastewater treatment
	HPZ/SZ of agricultural objects
	HPZ/SZ of agrochemical businesses
	HPZ/SZ of animal productions
	HPZ/SZ of store of manure and field airports
Transport and other structures	Protective zones (PZ) of railways
	PZ of roadways
	PZ of airports and landing path of airports
	PZ of ports
	PZ of electrical cables
	PZ of gas facilities
	PZ of gas and oil pipelines
	PZ of cables
	PZ of storages
	Military zones

Table 3.10 Frequently used typological characteristics of the SEF bound to the deterioration of the environment declared in acts and other planning and development documentation

Elements of primary and secondary landscape structure	SEF defined in various regulations
Sources of pollution	Stationary sources of atmospheric pollution • Particularly important technological unit ('large source') • Important technological unit, if not part of a large source ('medium source') • Other technological units ('small source')
	Mobile sources of air pollution
	Sources of water pollution (differed according to size)
	Noise sources (depending on size)
	Sources of radiation
Territories with air pollution	Area of air pollution • Area of air quality management • Defined parts of zones or agglomerations with an area of 50 km^2 • Where there occur odorous pollutants in concentrations which are nuisance to the population
	Area burdened by ground-level inversions
	Area burdened by noise (by degree)
Territories with water pollution	Sections of polluted water courses (by category)
	Area of polluted groundwater above the norm
	Sensitive areas
	Vulnerable areas
	Declared inundation area
Soil damage and soil with reduced conditions	Territory with contaminated soil
	Erosion of damaged soil (by degree)
Damaged vegetation	Damage to vegetation—defoliation (by degree)
	Forests damaged by emissions
Other areas threatened by stress factors	Environmental burden according to Register of environmental burdens A, B, C
	Area threatened by landslides
	Area threatened by avalanches
	Undermined area
	Area of increase seismicity
	Area of increased natural radioactivity

Table 3.11 Frequently used typological characteristics of the SEF bound to the administrative boundaries in the landscape

Elements of primary and secondary landscape structure	SEF defined in various regulations
Territory as a whole	**General Territorial Division**
	Administrative boundaries (state, district, municipalities)
	Cadastre boundaries
	Boundaries of urban areas and urban zones
	Sectoral boundaries
	Boundaries of landowner societies
	Boundaries of land-use planning units (territorial units, settlement units)
	Boundaries of forest enterprises
	Boundaries of agricultural enterprises
	River basin district, river basin organisations
	Boundaries of hunting and fishering grounds
	Declared areas of pheasantry

Required activity → decision-making → proposed area for activity → realisation → CLS element

Differentiating the required activities and existing CLS elements is particularly important in the decision-making process because:

- the required activities **may not yet exist**. Even in maps and plans, they are yet merely intangible socio-economic phenomena and represent a **request** for a change of CLS. Therefore, they must be assessed according to **their expected impact** on the landscape once implemented, i.e. as if they already existed in the proposed area;
- nevertheless, on each area already exists a ceratin CLS element presently. These are elements either to be retained, or to be changed. They provide knowledge of **an actual impact** of a particular element, as a basis for abstract decision-making.

Explanation on examples:

Required activity	Proposed area for required activity	CLS element after the realisation of the proposal
Erosion prevention	Landscape greenery	Shrubs
Plant production	Arable land	Crop field
Livestock production	Grassland	Pasture
Fruit production	Permanent culture	Orchard
Outdoor recreation	Recreational area	Camping
Housing	Built-up area	Dwelling houses

As we can see, the areas of particular activities and the CLS elements after the implementation are almost identical. For practical reasons in the LANDEP decision-making process, the required activities are referred to in the same way as the CLS elements—either the existing or the proposed ones. It is also appropriate to use the same names for the required activities as those that are used in the planning process for which LANDEP is being processed. For example, if LANDEP forms the basis for a territorial plan of a particular area, it is advisable to use identical names for the required activities with the names in that territorial plan.

The most frequently required social activities in landscape plans are categorized according to their nature, similarity in physical character, functions, dynamics, and according to their expected impact on the landscape as follows:

- ecostabilising activities (greenery, forests, water areas),
- permanent cultures (vineyards, orchards, gardens, gardens with cottages),
- agricultural activities (arable land with wide use, permanent grasslands),
- recreational activities (cottage areas, recreational areas),
- investment activities (diverse built-up areas).

Possible selection and sequencing of required activities are shown in Table 3.12.

Table 3.12 shows that from the physiognomic-ecological standpoint, the activities are listed from the most natural ecostabilising activities—greenery, forests—through the agricultural activities, recreation, housing, to the production and technical activities. Obviously, in each project different required activities may be selected; however, they should be sequenced in a similar way.

In the final step of LANDEP method—in the proposal for the ecologically optimal land use (see below)—the areas for the proposed activities are referred to as **functional types** and **functional regions** of the landscape.

Table 3.12 An example of selection of required social activities

Group of activities	Subgroup of activities	Activities	The label of area for the required activity
E			**Ecostabilizing activities**
	E1		*Non-woody tree and shrub vegetation—ecological greenery*
		E11	Greenery for soil and water erosion control
		E12	Greenery of climate and hygienic importance
	E2		*Forests*
		E21	Primaeval forests, forests without management
		E22	Special purpose forests
		E23	Protection forests (soil and water protection)
		E24	Commercial forests
		E25	Forest nurseries
		E26	Animal farms in the forest (farms, pheasantry)
	E3		*Wetlands and water areas*
		E31	Wetlands
		E32	Natural water areas
		E32	Natural water flows
		E33	Ponds
		E34	Tanks
		E35	Regulated water flows and channels
P			**Agricultural activities**
	P1		*Permanent grasslands*
		P11	Extensive pastures and meadows managed by mowing
		P12	Meadows managed by medium intensity mowing
		P13	Intensive pastures (fenced roads, paddock manuring of pastures)
		P14	Intensive meadows managed by mowing
	P2		*Arable land with big blocks*
		P21	Perennial forage crop production
		P23	Production of main crops on arable land
		P24	Production of non-food crops
		P25	Production of vegetable
		P26	Production of special crops on arable land
		P27	Greenhouse and plastic greenhouse management
	P3		*Arable land with small blocks*
		P31	Terraced fields
		P32	Arable land with small blocks in gardens and mosaic of different cultures

(continued)

Table 3.12 (continued)

Group of activities	Subgroup of activities	Activities	The label of area for the required activity
	P4		*Permanent cultures*
		P41	Vineyards
		P42	Orchards
		P43	Gardens
		P44	Hop gardens
R			**Recreational activities in the country**
	R1		*Extensive recreational activities*
		R11	Summer tourism
		R12	Winter tourism (skiing)
		R13	Temporary summer recreational areas (by the water)
		R14	Temporary winter recreational areas (ski slopes)
		R15	Camping sites, public camps
	R2		*Recreation in objects*
		R21	Country houses
		R22	Cottages, cottage settlements
	R3		*Recreation*
		R31	Sports and recreational facilities (playgrounds, stadiums)
		R32	Recreational centers
		R33	hotels
B			**Investment activities—housing and civil construction**
	B1		*Housing construction*
		B11	Individual housing construction
		B12	Complex housing construction
		B13	Service
V			**Investment activities—industry and mining**
	V1		*Industry*
		V11	Warehouses
		V12	Craft production, small-scale industry
		V13	Food industry
		V14	Consumer, light and engineering industry
		V15	Metallurgical and energy industry
		V16	Heavy, chemical and wood-chemical industry
	V2		*Mining for mineral resources*
		V21	Surface mining of minerals
		V22	Underground mining of minerals

(continued)

Table 3.12 (continued)

Group of activities	Subgroup of activities	Activities	The label of area for the required activity
T			**Agricultural–technical objects**
	T1		*Machine–tractor stations*
	T2		*Animal production objects*
D			**Traffic objects and areas**
	D1		*Objects of car transport*
		D11	Highways
		D12	Different types of roads
	D2		*Objects of railway transport system*
		D21	Train station objects
		D22	Railway track
	D3		*Airports*
S			**Waste management, landfills, sludge beds**
	S1		*Ponds, landfill sites*
	S2		*Landfills (municipal* solid waste*)*
I			**Objects of infrastructure**
	I1		*Different technical objects*

References

Bandrova T, Konečný M, Zlatanova S (2010) Geographic information and cartography for risk and crisis management. Towards better solutions. Lecture notes in geoinformation and cartography. Springer, London

Bedrna Z et al (1992) Analýzy a čiastkové syntézy zložiek krajinnej štruktúry. Slovenská technická knižnica, Bratislava

Bedrna Z (2002) Environmentálne pôdoznalectvo. VEDA, Bratislava

Burrough PA, McDonnel RA (1998) Principles of geographical information systems. Oxford University Press, Oxford

Csorba P, Szabó Sz (2012) The application of landscape indices in landscape ecology. In: Tiefenbacher J (ed) Perspectives on nature conservation—patterns, pressures and prospects. InTech, pp 121–140. Available from: http://www.intechopen.com/books/perspectives-on-natureconservation-patterns-pressures-and-prospects/the-application-of-landscape-indices-in-landscape-ecology

Džatko M et al (1976) Charakteristika bonitovaných pôdno-ekologických jednotike SSR. MPVž, Bratislava

Herzog F, Lausch A (1999) Prospects and limitations of the application of landscape metrics for landscape monitoring. In: Maudsley M, Marshall J (eds) Heterogeneity in landscape ecology: pattern and scale. IALE(UK), Aberdeen, pp 41–50

Hraško J et al (1991) Morfogeneticky klasifikačný systém pôd ČSFR [The morfphogenetic classification system of the soils of CSFR], 2nd edn. Výskumný ústav pôdnej úrodnosti VÚPÚ [Research Institute for soil fertility], Bratislava

Izakovičová Z et al (2000) Metodické pokyny na vypracovanie projektov regionálnych ÚSES a miestnych ÚSES. Združenie krajiny 21, Bratislava

Kocsis K, Agárdi N, Koczó F, Mezei GA, Nemerkényi Zs, Szabó R (2016b) Generalizálás géppel, vagy kézzel? - módszertani összehasonlítás Magyarország Nemzeti Atlasza példáján. In: Pajtókné Tari I, Tóth A (ed) Magyar Földrajzi Napok: Absztraktkötet: VIII. Magyar Földrajzi Konferencia. Magyar Földrajzi Társaság, Budapest, p 103

Konečný M, Kubíček P, Duda P, Herman L, Kolář M, Kozel J, Mulíčková M, Řezník T, Stachoň Z, Štampach R, Štěrba Z (2012) Datové infrastruktury pro prostorově informační společnost. Masarykova univerzita, Brno

Miklós L et al (1986) Ekologický plán VSN. Súbor grafických výstupov. Ekologická optimalizácia využívania VSN. IV. diel. ÚEBE SAV Bratislava - Slovosivo

Miklós L, Izakovičová Z (1997) Krajina ako geosystém. Veda SAV, Bratislava

Miklós L, Špinerová A (2011) Krajinno-ekologické plánovanie LANDEP [The landscape-ecological planning LANDEP]. VKU, Harmanec

Ružička M, Miklós L (1982b) Landscape-ecological planning (LANDEP) in the process of territorial planning. Ekológia (ČSSR) 1:297–312

Ružička M, Miklós L (1990) Basic premises and methods in landscape-ecological planning and optimisation. In: Zonnenveld IS, Forman RTT (ed) Changing landscapes: an ecological perspectives. Springer, New York, pp 233–260

Ružičková H, Ružička M (1973) Druhotná štruktúra krajiny ako kritérium biologickej rovnováhy. Quaestiones geobiologicae 12:23–62

Špinerová A (2005) Charakteristika vybraných vlastností mikropovodí v povodí Ilijského potoka. Rigorózna práca. Univerzita Mateja Bela, Katedra krajinnej ekológie FPrV, Banská Bystrica

Špinerová A (2010) Krajinno-ekologické limity poľnohospodárskeho využitia Ilíjskeho potoka. VKÚ, a.s., Harmanec

Špinerová A (2011) Prvky využitia zeme ako regulátor zón dynamiky svahu v modelovom území Ilijského potoka. Doktorská dizertačná práca. TU, Zvolen

Šúriová N, Izakovičová Z (1995) Territorial system of anthropogenic stress factors in landscape ecological planning. Ekológia (Bratislava) 14(2):181–189

Szabó SZ, Csorba P (2009) Tájmetriai mutatók kiválasztásának lehetséges módszertana egy esettanulmány példáján. Tájökológiai Lapok 7(1):141–153

Verrasztó Z (1979) Land formation and the geological aspects of environmental protection. In: Symposium changes of the geological environment under the influence of man's activity. IAEG National group, Krakow-SandomierzBelchatow-Plock-Warszawa, pp 135–141

Verrasztó Z (2010) Környezeti monitoring vizsgálatok az Ipoly vízgyűjtőjén. Tájökológiai Lapok 8(3):532–561

Chapter 4
Landscape-ecological Syntheses in LANDEP

Abstract Syntheses are aimed to create a basis for right decision-making of landscape, which are given as follows:

- knowledge on the values of **all** indicators of the properties of geocomplexes on **every single spot** of the territory,
- **demarcation** of homogeneous areas with a defined combination of values—mapping of homogenous geocomplexes.

Accordingly, the syntheses in the LANDEP methodics are defined as follows:

- formation, description and classification of geocomplexes of various ranges according to a clearly defined set of analytical properties, indicators and values,
- the spatial delineation of homogenous areas of geocomplexes according to these values and,
- the segmentation of the territory into these homogeneous areas.

Geocomplexes with various characteristics and values of indicators—most often the abiotic complexes and landscape-ecological complexes—are then carriers of integrated information about the landscape.
The most frequently used method of spatial synthesis in LANDEP is the superposition of homogenised analytical maps in defined order of leading indicators. The result of such syntheses—marked also as topical syntheses—are the **types** of abiotic complexes and landscape-ecological complexes. Hierarchically higher synthetic units—regions—are created on the base of the spatial structure and content of lower hierarchical units. The chapter presents few examples of choric syntheses too.
A substantial step of the syntheses is the creation of the geosystem-based classification of geocomplexes formally arranged to the tables of abiocomplexes and landscape-ecological complexes.
The syntheses are the key step of LANDEP procedure. For the further steps of LANDEP, the synthetic spatial landscape-ecological units—the types of geocomplexes of different character —constitute the following:

L. Miklós and A. Špinerová, *Landscape-ecological Planning LANDEP*,
https://doi.org/10.1007/978-3-319-94021-2_4

- a comprehensive system of information on the landscape, object-attribute spatial database;
- the basis for the interpretations—the formation of the criteria (utility properties) for the evaluation process of the landscape;
- the object of evaluation as regards to selected human activities;
- the boundaries of the areas for the proposals for the optimal land use, therefore they are the subject of proposals.

The chapter is illustrated by numerous tables and maps which document also the historical development of the LANDEP methodics.

Keywords Complexes · Superposition · Types · Regions · Databaes

A basic requirement for the right decisions on the utilisation of each spot of the landscape is predetermined by the correct information on the synergic effect of the values of property indicators of geocomplexes. Therefore, a prerequisite for the right decision-making on land use is as follows:

- knowledge on the values of all indicators of the properties of the geocomplexes **on each spot**—in other words, to know the combination of these values of geocomplexes;
- **demarcation** of homogeneous areas with a defined combination—mapping of homogenous geocomplexes.

This aspect of the landscape-ecological planning is solved in the methodological step of landscape-ecological synthesis. The aim of synthesis is the logical arrangement—in some cases, partial generalisation—of a set of indicators of the properties of the geocomplexes so that they allow the creation of clearly and simply characterised spatial units with the **complex** set of values of analysed indicators of properties, i.e. characteristics and demarcation of areas of spatial units of landscape—**geocomplexes** of different order and different content.

Certain combinations of values of the indicators of a geocomplex have their own spatial distribution and create **homogeneous areas**, distinct from adjacent homogeneous areas with different combinations of values. The areas of geocomplexes can be depicted on a map (mapped). They divide the territory without remainder.

Geocomplexes with various characteristics and values of indicators, hierarchical levels and types according to their internal structure are then carriers of integrated information about the landscape; thus, they create the real **landscape-ecological basis for integrated management**.

The step of "synthesis" in the LANDEP methodics is defined as

- formation, description and classification of geocomplexes of various ranges on the basis of a clearly defined set of analytical properties, indicators and values,
- the delineation of homogenous areas of geocomplexes according to these values and
- the segmentation of the territory into these areas of geocomplexes.

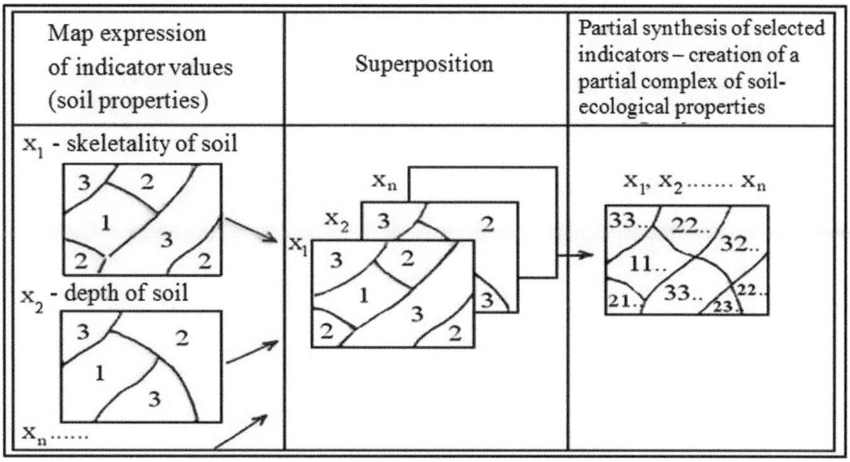

Fig. 4.1 Scheme of the spatial synthesis by superposition of analytical materials

4.1 Topical Synthesis—Formation of Topical Geocomplexes (Abiocomplexes and Landscape-ecological Complexes)

The most commonly used method of syntheses is the so-called method of superposition of analytical maps, resulting in the parametric typisation of the territory—creating maps of the types of abiotic complexes ABC, biotic complexes BIC, socio-economic complex SEC or landscape-ecological complexes LEC. Currently, the technology of geographical information systems (GIS) is being used.

There are different techniques of spatial parametrical landscape-ecological syntheses (Džatko et al. 1976; Mičian 1982; Hreško et al. 2008). The logic of this procedure is very simple (Fig. 4.1).

Geocomplexes of various range and complexity represent the topical model of synthetic units, where model elements are the indicators of their properties $\boldsymbol{x_n}$ (state variables or spatial subsystems), generally displayed as

$$G_K(x_1, x_2, \ldots x_n).$$

According to this scheme, various types of geocomplexes are created for the purpose of LANDEP. The most frequent procedures are as follows:

4.1.1 Formation of Abiotic Complexes ABC

The above-mentioned simple method of syntheses by means of superposition of analytical maps is mainly used in the formation of abiotic complexes **ABC**. We illustrate the process based on the example of the formation of abiotic complexes on Ilijský stream catchment (Špinerová 2010).

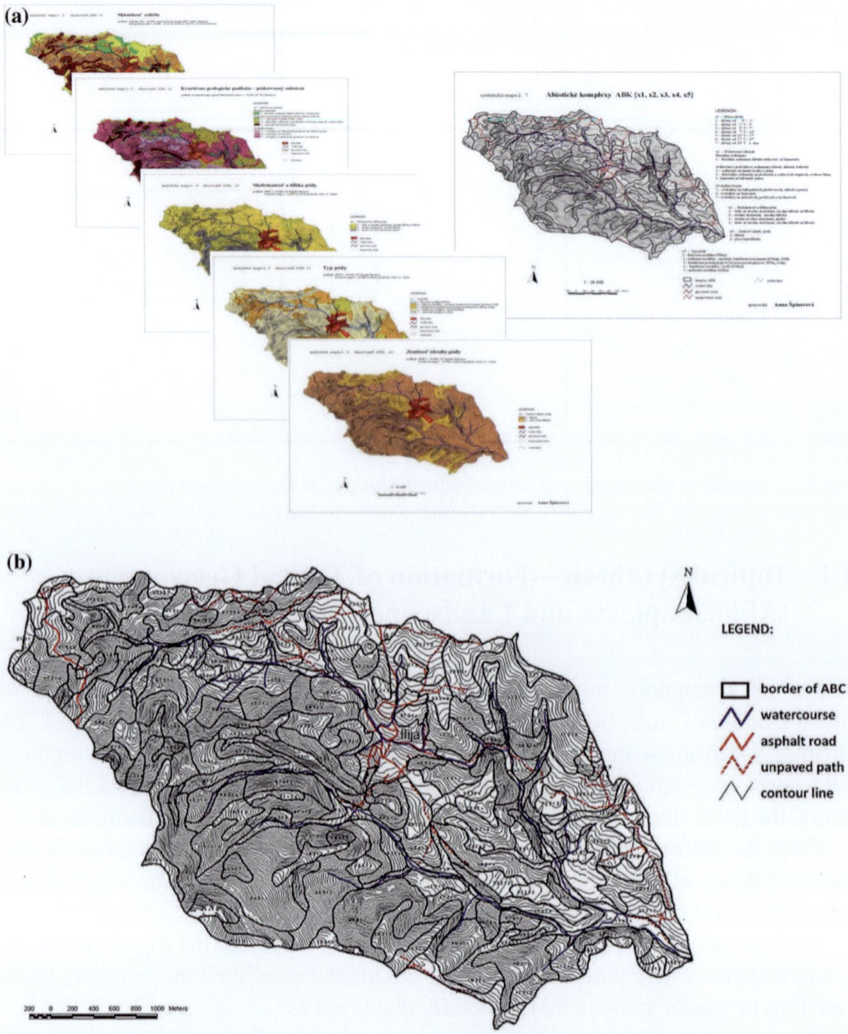

Fig. 4.2 **a** Illustration of the process of creation of the types of abiotic complexes ABC. The maps on the left side: the analytic maps—inputs to the synthesis—illustrated in Figs. 3.3–3.7. The map on the right side: the synthetic map of abiocomplexes ABC as a result of synthesis—illustrated in Fig. 4.2b. **b** The result of the creation of the types of abiotic complexes ABC: **ABC (x_1, x_2, x_3, x_4, x_5)**. Each polygon represents area of ABC with homogenous values of all five included indicators of the properties (described with five codes in electronic form). The included indicators: $\mathbf{x_1}$—Slope angle, $\mathbf{x_2}$—Geological/soil-forming substrate, $\mathbf{x_3}$—Soil skeletality and depth, $\mathbf{x_4}$—Soil texture, $\mathbf{x_5}$—Soil type

Table 4.1 The methodical frame for the creation of the Table of types of ABC

Types of ABC				
Analytical indicators and their values				Abiotic complex ABC
x_1—inclination	x_2—soil depth	x_3—…	x_n—…	$(x_1, x_2, x_3, \ldots x_n)$
1. Code (value)	2. Code (value)			Code of type of ABC
1 (0°–1°)	**1** (deep)	**…**	**…**	**11…n**
	2 (medium)			**12…n**
	3 (shallow)			**13…n**
	4 (no soil)			**14…n**
2 (1°–3°)	**1** (deep)			**21…n**
	…			**….**
	…			**….**

The following indicators of the properties of primary landscape structure were included in the process of synthesis: $\boldsymbol{x_1}$—relief inclination, $\boldsymbol{x_2}$—soil-forming substrate, $\boldsymbol{x_3}$—skeletality and soil depth, $\boldsymbol{x_4}$—soil texture and $\boldsymbol{x_5}$—soil type. There are many analytical indicators in each territory, but not all differentiate the given territory, or for given purposes are not relevant. In the case study the selected indicators were the most important, the others did not give much more practical information.

The system description of the abiotic complex in this example is as follows:

$$\boldsymbol{ABC}\,(\boldsymbol{x}_1, \boldsymbol{x}_2, \boldsymbol{x}_3, \boldsymbol{x}_4, \boldsymbol{x}_5)$$

The process of the formation of ABC in the model area is shown in Fig. 4.2a, the result of the Fig. 4.2b. The analyses and syntheses were processed in GIS, the displayed figures should serve just as a demonstration of the process.

Areas with the same combination of values of indicators of ABC may occur repeatedly in a sufficiently large territory; therefore, they create typological units marked as Types of Abiotic Complexes ABC.

Combinations of indicator values should be arranged in a simple Table of Types of ABC (Table 4.1). Despite the simplicity of the table, its logical structure must be maintained, assuming it will be the basis for other functional tables, as the table of the types of landscape-ecological complexes, as well as the decision table in the next steps. The system header of the ABC table in whatever territory and with whatever set of indicators $\boldsymbol{x_n}$ should keep the following form:

Table 4.2 shows an example of the types of ABC that are described by a set of five codes on the model area of Ilijský creek catchment area.

4.1.1.1 Methodological Notes on the Formation of Abiotic Complexes

The landscape-ecological syntheses do not merely mean mechanical superposition of maps; they represent the actual relations within the elements of geosystems. Therefore, one should realise the following:

Table 4.2 Example of the table of types of abiotic complexes ABC (Ilija creak catchment)

No.	Slope angle x_1	Substratum x_2	Skeletality and depth of soils x_3	Soil texture x_4	Soil types x_5	Code of the types of ABC (x_1, x_2, x_3, x_4, x_5)	Number of appearance in the model territory
1.	0	0	0	0	0	0, 0, 0, 0, 0	1
2.	1	1	3	1	1	1, 1, 3, 1, 1	32
3.	2	1	3	1	1	2, 1, 3, 1, 1	16
4.	2	2	2	1	2	2, 2, 2, 1, 2	1
5.	2	2	2	1	3	2, 2, 2, 1, 3	18
6.	2	3	2	1	2	2, 3, 2, 1, 2	7
7.	2	3	2	1	3	2, 3, 2, 1, 3	5
8.	2	3	2	1	4	2, 3, 2, 1, 4	1
9.	2	3	3	2	4	2, 3, 3, 2, 4	2
10.	2	7	3	1	4	2, 7, 3, 1, 4	4
11.	3	2	2	1	2	3, 2, 2, 1, 2	27
12.	3	2	2	1	3	3, 2, 2, 1, 3	21
13.	3	2	2	1	4	3, 2, 2, 1, 4	1
14.	3	2	3	1	2	3, 2, 3, 1, 2	1
15.	3	3	2	1	2	3, 3, 2, 1, 2	55
16.	3	3	2	1	3	3, 3, 2, 1, 3	6
17.	3	3	2	2	2	3, 3, 2, 2, 2	11
18.	3	3	2	2	3	3, 3, 2, 2, 3	6
19.	3	3	3	2	4	3, 3, 3, 2, 4	7
20.	3	3	4	1	4	3, 3, 4, 1, 4	1
21.	3	5	2	1	2	3, 5, 2, 1, 2	4
22.	3	6	3	2	4	3, 6, 3, 2, 4	5
23.	3	7	2	1	4	3, 7, 2, 1, 4	2
24.	3	7	3	1	4	3, 7, 3, 1, 4	8
25.	4	4	3	1	4	4, 4, 3, 1, 4	1
26.	4	5	2	1	2	4, 5, 2, 1, 2	13
27.	4	5	2	1	4	4, 5, 2, 1, 4	7
28.	4	5	2	2	4	4, 5, 2, 2, 4	22
29.	4	6	2	1	4	4, 6, 2, 1, 4	1
30.	4	6	3	1	2	4, 6, 3, 1, 2	4
31.	4	6	3	1	4	4, 6, 3, 1, 4	9
32.	4	6	3	2	4	4, 6, 3, 2, 4	10
33.	4	7	2	1	2	4, 7, 2, 1, 2	37
34.	4	7	2	1	4	4, 7, 2, 1, 4	44
35.	4	7	2	2	4	4, 7, 2, 2, 4	19
36.	5	4	3	1	4	5, 4, 3, 1, 4	2
37.	5	4	4	1	4	5, 4, 4, 1, 4	1
38.	5	5	2	2	4	5, 5, 2, 2, 4	2
39.	5	6	3	1	2	5, 6, 3, 1, 2	4
40.	5	6	3	1	4	5, 6, 3, 1, 4	4
41.	5	6	3	2	4	5, 6, 3, 2, 4	15

(continued)

Table 4.2 (continued)

No.	Slope angle x_1	Substratum x_2	Skeletality and depth of soils x_3	Soil texture x_4	Soil types x_5	Code of the types of ABC (x_1, x_2, x_3, x_4, x_5)	Number of appearance in the model territory
42.	5	7	2	1	2	5, 7, 2, 1, 2	8
43.	5	7	2	1	4	5, 7, 2, 1, 4	37
44.	5	7	2	2	4	5, 7, 2, 2, 4	10
45.	6	4	3	1	2	6, 4, 3, 1, 2	4
46.	6	4	3	1	4	6, 4, 3, 1, 4	9
47.	6	4	3	2	4	6, 4, 3, 2, 4	5
48.	6	4	4	1	4	6, 4, 4, 1, 4	4
50.	6	5	2	1	2	6, 5, 2, 1, 2	8
51.	6	5	2	1	4	6, 5, 2, 1, 4	3
52.	6	6	3	1	2	6, 6, 3, 1, 2	2
53.	6	6	3	2	4	6, 6, 3, 2, 4	7
54.	6	6	4	1	4	6, 6, 4, 1, 4	2
55.	6	7	2	1	4	6, 7, 2, 1, 4	4
56.	6	7	2	1	5	6, 7, 2, 1, 5	1
57.	6	7	3	1	4	6, 7, 3, 1, 4	11
58.	6	7	4	1	2	6, 7, 4, 1, 2	1
59.	6	7	4	1	4	6, 7, 4, 1, 4	6
60.	7	4	4	1	2	7, 4, 4, 1, 2	2
61.	7	4	4	1	4	7, 4, 4, 1, 4	3
62.	7	6	4	1	2	7, 6, 4, 1, 2	1
63.	7	7	3	1	4	7, 7, 3, 1, 4	1

Number of types of ABC = 63
Number of areas of ABC = 567
The codes in individual columns present the value of the related indicator x_n on the area of ABC. In the column, "*Code of the types of ABC*" is presented the combination of values of each indicator on the area of given type of ABC

- some combinations of the values of the single indicators of ABC, or combinations of the types of ABC with the types elements of current landscape structure CLS are completely impossible or at least unrealistic.
- obviously, it cannot be anticipated in advance what sort of impossible, unrealistic or unlikely combinations of values may occur in the model area.

A guide to a critical evaluation of combinations of the most commonly occurring indicators are provided in the next tables.

The Table 4.3 illustrates the characteristics of the combinations of morphological-morphometric relief types versus genetic–lithological substrate types; the Table 4.4, the combination of genetic–lithological substrate types versus soil association types and the Table 4.5 the combinations of ABC versus CLS.

Table 4.3 The possibilities of the mutual combinations of the values of indicators of morphologic-morphometric types of relief versus genetic–lithologic types of the substratum

Geology.			*Fluvial*		*Slope*		*Aeolic*				*Deluvial*			*Glacial*		*Organic*		*Anthropic*		*Eluvial*
Morphology ↓	Verbal characterics	Inclination	Alluviums	Terrace gravels	Koluvium	Proluvium	Loess	Loess loams	Polygenetic loams	Sands	Loams	Screes	Landslides	Moraines	Glacial sediments	Fens	Peats	Mounds	Quarries	Rocks
Top	peak	ext																		
	top	3-7(12)																		
Ridge	flat	1-3																		
	vaulted	3-7																		
Crest	flat	1-3																		
	vaulted	3-5																		
	slightly rugged	5-7(12)																		
Plateau	top plateau and tableland	(1)3-7																		
Slope plateau, slope step	river and terrace steps	(1)-3-(5)																		
Slope	very mild	1-3																		
	mild	3-7																		
	middle	7-12																		
	steep	12-17																		
	very steep	17-25																		
	extremely	25-35																		
	scarp	≥ 35																		
Piedmont	terrace, alluvial cone	1-5(7)																		
Valley and dell bottom	valley and dell bottom	1-3																		
Plain	alluvium	0-1																		
Wavy plain	lowland plain alluvium and terrace	0 - 1																		
	loess table	1-3																		
	sand dune	1-3 (5)																		
	aggradation mound, meander plain	0-1(3-(5)																		
Depression	recess, lowland depression	0-(1)																		

Table 4.4 The possibilities of the mutual combination of the values of indicators of the types of geological substratum versus soil association types

Geological substratum →	*Fluvial*		*Slope*		*Aeolic*				*Deluvial*			*Glacial*		*Organic*		*Anthropic*		*Eluvial*
Soils ↓	Alluviums	Terrace gravels	Koluvium	Proluvium	Loess	Loess loams	Polygenetic loams	Sands	Loams	Screes	Landslides	Moraines	Glacial sediments	Fens	Peats	Mounds	Quarries	Rocks
Gleysol																		
Fluvisol																		
Mollic fluvisols																		
Chernozem																		
Luvisol typical																		
Albic Luvisol																		
Pseudogley																		
Cambisol																		
Podzol																		
Pararendzic Leptosol																		
Rendzic Leptosol																		
Regosol																		
Organosol																		
Leptosol																		
Anthrosol																		
Without soil																		

Table 4.5 The methodical frame for the creation of combination table—Table of types of ABC and CLS elements

PLS – Types of ABC ***($x_1, x_2, ..., x_n$)***				CLS – Current landscape structure						
x_1 - slope	x_2 – soil depth	x_n -	Code of ABC type	Code of CLS						
				1	2	3	4	5	6	7
1st code - *(value)*	2nd code *(value)*	x[th] code		**Forests**	**Greenary**	**Permanent cultures**	**Permanent grasslands**	**Arable land**	**Built-up areas**	**Water areas**
1 *(0-1°)*	**1** *(deep)*		**11...**	11...1	11...2		11...4	11...5	11...6	
	2 *(moderate)*		**12...**	12...1	12...2	12...3	12...4	12...5	12...6	12...7
	3 *(shallow)*		**13...**	13...1						
	4 *(without soil)*		**14...**		14...2					
2 *(1-3°)*	**1** *(deep)*		**21...**					21...5		
	2 *(moderate)*		**22...**		22...2	22...3	22...4	22...5		
	3 *(shallow)*		**23...**		231				23...6	23...7
	4 *(without soil)*		**24...**							
3 *(3-7°)*	**1** *(deep)*		**31...**			31...3		31...5		
	2 *(moderate)*		**32...**			32...3	32...4	32...5		
	3 *(shallow)*		**33...**	33...1	33...2		33...4		33...6	33...7
	4 *(without soil)*		**34...**		34...2				34...6	34...7
4 *(7-12°)*	**1** *(deep)*		**41...**			41...3	41...4		41...6	
	2 *(moderate)*		**42...**	42...1		42...3	42...4		42...6	
	3 *(shallow)*		**43...**		43...2					
	4 *(without soil)*		**44...**		44...2					
5 *(12-17°)*	**1** *(deep)*		**51...**	51...1	51...2					
	2 *(moderate)*		**52...**	52...1	52...2		52...4			
	3 *(shallow)*		**53...**		53...2					
	4 *(without soil)*		**54...**		54...2					
6 *(17-25°)*	**1** *(deep)*		**61...**							
	2 *(moderate)*		**62...**	62...1	62...2		62...4			
	3 *(shallow)*		**63...**		63...2		63...4			
	4 *(without soil)*		**64...**							
7 *(>25°)*	**1** *(deep)*		**71...**							
	2 *(moderate)*		**72...**	72...1	72...2					
	3 *(shallow)*		**73...**		73...2					
	4 *(bez pôdy)*		**74...**							

Note The table is an illustration of the methodic step. Its content is fictive, not bound to concrete territory

The occurring combinations of LEC types in the squares are marked with combination of codes of types of ABC and code of CLS

The red and orange colours mark the impossible or unlikely combination, grey colour marks the combinations not occurring in the model territory

The scale of possibilities of occurrence of combinations of indicator values in Tables 4.3 and 4.4 are illustrated with the appropriate colouring, as follows:

Most characteristic combinations	
Typical	
Likely, possible	
Possible in specific cases	
Neutral but rare	
Unlikely, thorough examination required	
Impossible	

If in the process of syntheses such combinations occur, which are characterised as impossible or unlike, there has certainly been an error, either in the analytical maps or in the process of syntheses. These combinations have to be corrected!

4.1.2 *Formation of Landscape-ecological Complexes LEC*

Next step of the landscape-ecological synthesis is the synthesis of abiotic with biotic–anthropogenic complexes that, according to the theoretical–methodological principles of the assessment of the current landscape structure (Chap. 3.3), are referred to as mapping units **CLS** elements. For our purposes, we marked the CLS elements as $\boldsymbol{y_1}$.

Synthesis of ABC types and CLS elements results in more complex units: **landscape-ecological complexes LEC**.

The system entry of a landscape-ecological complex is

$$\boldsymbol{LEC\{(ABC), y_1\}},$$

more precisely,

$$\boldsymbol{LEC\{ABC(x_1, x_2, x_3, x_4, x_5), y_1\}}$$

Areas with the same combination of values of analytical properties represent typological units—**types of LEC**. In the case of LEC, the combinations of values should also be arranged in a table. The methodological scheme of the ABC/CLS combination is in Table 4.5. It shows a scheme of simplified example, where ABC types are characterised by only two most commonly used indicators: slope inclination (eight classes of values) and soil depth (four classes of values). As for example, the CLS is characterised by seven classes of values.

The occurring combinations of LEC types in the squares are marked with combination of codes of types of ABC and code of CLS.

The grey colour marks other combinations which do not occurr in the model territory. The red and orange colours mark the impossible or unlikely combination. The explanation is as follows:

(grey)	• Combinations not occuring in the model territory
(red)	• Impossible or ireal mutual combinations of the elements of ABC as e.g. • ABC on slopes over 25° inclination with deep soils • Impossible or ireal combinations of the elements of ABC with elements of CLS, as e.g. ➢ ABC defined as "without soil" but with arable land ➢ ABC over 25° with arable land or other intensive cultures ➢ ABC over 7° with water surfaces
(orange)	• Unlikely murual combinations of the elements of ABC as e.g. ABC over 17° with deep soils • Unlikely combinations of the elements of ABC with elements of CLS, as e.g ABC defined as „without soils „ but with productive forests, arable land an other intensive cultures • ABC over 17° with arable land or with dwelling houses

Of course, there are many other possible unlikely combinations. Nevertheless, if such combinations occur, a deeper analysis of possible mistakes is needed.

A real example of the formation of types of LEC is illustrated on the Ilijský creek catchment area. The process of a complex synthesis involved real abiotic complexes ABC (as described in Table 4.1 and the elements $\boldsymbol{y}_1$—current landscape structure CLS (Table 3.6). The real combinations of types of ABC and CLS are shown in Table 4.6. Many theoretical combinations of the ABC and CLS, which are possible in the table, are not present in the real territory. These boxes of the table are not filled in.

Therefore Types of LEC in Table 4.6 are coded as follows:

- types of ABC characterised by five codes of the values of five analytical indicators $\boldsymbol{x}_1$–$\boldsymbol{x}_5$ based on Table 4.3 (the very left column),
- the actual combinations of ABC and CLS are indicated in the filled-in cells of the table, where the numbers represent their area in m^2 in the model territory.

An example of the creation and the result—the spatial distribution of the types of LEC on the model territory of the Ilijský screek catchment area as shown in Fig. 4.3a, b.

The analyses and syntheses were processed in GIS, the displayed version serves just as a demonstration of the process.

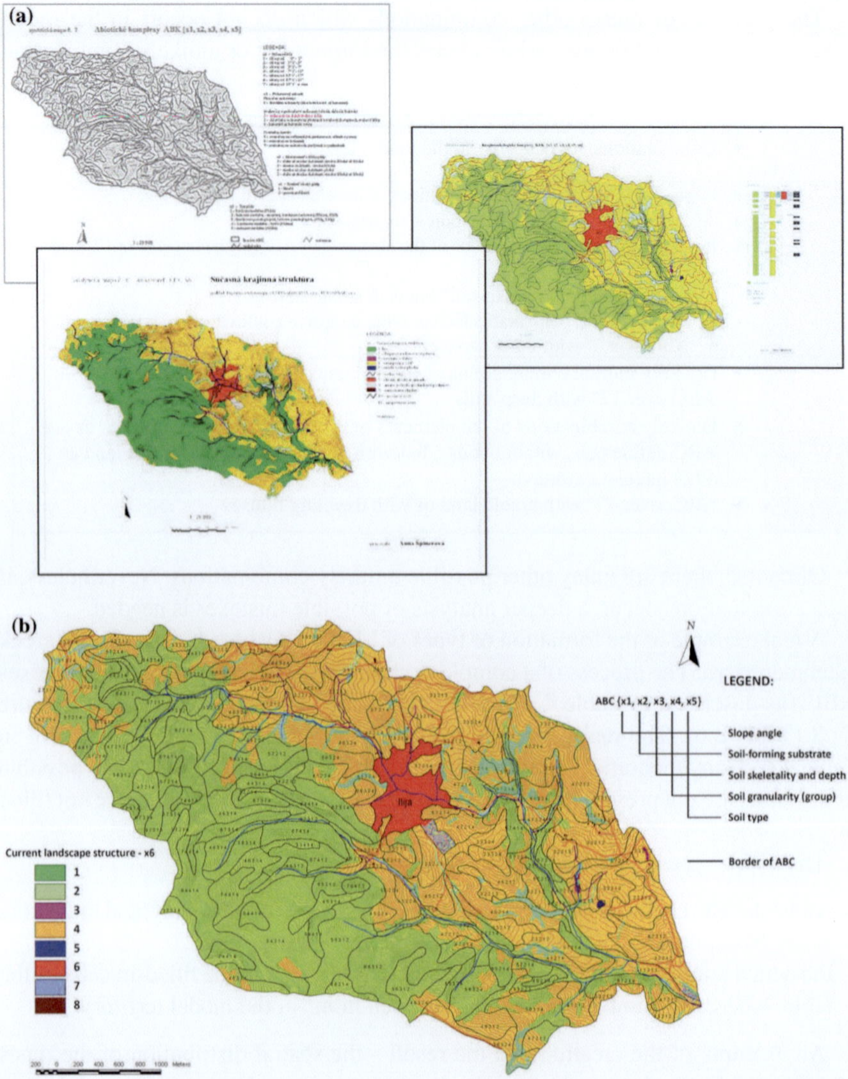

Fig. 4.3 **a** Illustration of the process of creation of the types of landscape-ecological complexes LEC. Maps on left side: the map of abiocomplexes ABC as shown in Fig. 4.2b and map of the current landscape structure CLS—reduced content of the original mapping shown in Fig. 3.8. Map on right side: the synthetic map of landscape-ecological complexes. See also Fig. 4.3b. **b** The result of the creation of the types of landscape-ecological complexes **LEC LEC** {**ABC** ($\mathbf{x_1}, \mathbf{x_2}, \mathbf{x_3}, \mathbf{x_4}, \mathbf{x_5}$), $\mathbf{x_6}$}, The abiotic complexes are characterised by indicators $\mathbf{x_1}$–$\mathbf{x_5}$ coded by five codes identically as in the Fig. 4.2b. The CLS elements $\mathbf{x_6}$ is illustrated by colours as: 1—Forest, 2—scrubs and linear vegetation, 3—marsh, fen, 4—arable land and permanent grassland, 5—water reservoir and creeks, 6—residential area and garden, 7—agricultural objects, 8—other objects (generalised picture with reduced number of CLS elements)

Table 4.6 Combination table of the types of ABC and CLS: the table of the types of landscape-ecological complexes LEC (Example on the Ilijský creek catchment area. Cut of the upper part of the complete list)

	Elements of CLS The elements with minor area were incorporated to neighbouring areas							
Code of the types of ABC	Forests	Scrubs	Meadows, marshes, muds	Arable land	Water surfaces	Residential areas	Agricultural objects	Other objects
0 0 0 0 0						382,491.80	44,851.10	
1 1 3 1 1	127,647.30	47,128.69		183,034.30			3868.04	1,102.89
2 1 3 1 1	103,046.00	29,243.91		20,601.98				
2 2 2 1 2				120,753.80				
2 2 2 1 3	25,665.15	5,438.70	3247.75	239,075.50	3097.39			227.03
2 3 2 1 2	761.55	13,252.63		166,211.50				782.38
2 3 2 1 3		4,009.60	2,516.91	119,024.60				
2 3 2 1 4				28,316.44				
2 3 3 2 4	5536.40			58,416.41				
2 7 3 1 4		2,902.96		42,676.88				
3 2 2 1 2	159,638.70	10,125.36		210,563.90			10,012.39	
3 2 2 1 3	54,917.18	26,410.72		158,864.20				836.60
3 2 2 1 4	12,851.45							
3 2 3 1 2	50,668.80							
3 3 2 1 2	44,135.07	28,494.27	2468.82	1,233,245.00	277.89			2295.27
3 3 2 1 3	2,968.94	4,729.57		219,673.20				479.34
3 3 2 2 2	8614.82	7819.98		103,246.70				
3 3 2 2 3	24,745.08	6,463.43		96,268.18				
3 3 3 2 4	5,210.23	2,527.43		258,280.50				554.44
3 3 4 1 4	36,820.46							
3 5 2 1 2	3067.36	1906.49		29,734.69				
3 6 3 2 4	2,092.68	20,046.67		107,972.70				
3 7 2 1 4	53,179.66			46,439.51				
3 7 3 1 4	59,437.47	24,631.43		66,720.69				
4 4 3 1 4	37,286.65							
4 5 2 1 2	22,784.44	12,852.07		221,096.10				169.11
4 5 2 1 4	7001.54	19,962.60		165,574.90				
4 5 2 2 4	180,278.30	35,318.44		254,753.30				360.76
4 6 2 1 4	26,906.07							

4.2 Choric–Spatial–Structural Synthesis—Formation of Complex Regions

There are many different well-developed approaches to choric syntheses. They represented one of the main streams of geographical researches where the main methodological problem was the determination of the borders of regions (Armand 1964; Bailey 2002; Kocsis and Hevesi 2003; Csorba 2008; Łowicki and Mizgajski 2013). The LANDEP methodology mostly uses syntheses aimed at creating hierarchically higher synthetic units—regions—on the base of the characteristic spatial structure and content of lower hierarchical units, in our case ABC and LEC types.

For illustration, we present four different approaches of the formation of higher units from different time periods, which are given as follows:

- Figure 4.4 shows a traditional physical–geographical typisation and regionalization based on leading typisation factors in the area of Gemerské Turce, year 1978 (Miklós 1978). The project aimed at the optimum recreational use of the model territory.
- Figure 4.5 shows a traditional division of the territory into geosystems based on the key integration factor—mesorelief—in the territory of the Eastern Slovak Lowland, year 1985 (Miklós et al. 1986). The project aimed at the optimalisation of the agricultural use of the eastern slovakian lowland.
- Figure 4.6 represents a parametric typing and regionalization of the territory based on alternating topical (vertical) and choric (horizontal) indicators in the Eastern Slovak Lowland, year 1986. The result is the division of the territory into geosystems of different types and different hierarchical levels (Miklós et al. 1986).
- Figure 4.7 illustrates a spatial synthesis of homogeneous run-off areas—microcatchments—into hydrological units of various levels in the Ilijský creek catchment area 2005 (Špinerová 2005). The project was aimed at the optimisation of the agricultural use of the catchment area and at slowing down the run-off from the territory.

4.3 Methodical Notes to the Landscape-ecological Syntheses

Within their borders, ABC and LEC have unchanging (homogeneous) values for all the selected indicators $\boldsymbol{x_n}$. Homogeneous areas of the types of landscape-ecological complexes (LEC) constitute the **spatial database** for the next steps of the LANDEP methodology. A set of values of the analytical properties of a given LEC type represents certain suitability for its use by man—i.e. a set of landscape-ecological conditions for deciding the optimal utilisation of the landscape—in its entire homogeneous area and on all other areas of occurrence of the given type. The synthetic units thus become the basic **operational units** of the subsequent decision-making process.

Synthesis is a key step in the LANDEP methodology. It is fair to say that it is also the final step in the basic research on the landscape, as well as the initial step for the applied procedures and planning practice.

The synthetic units are as follows:

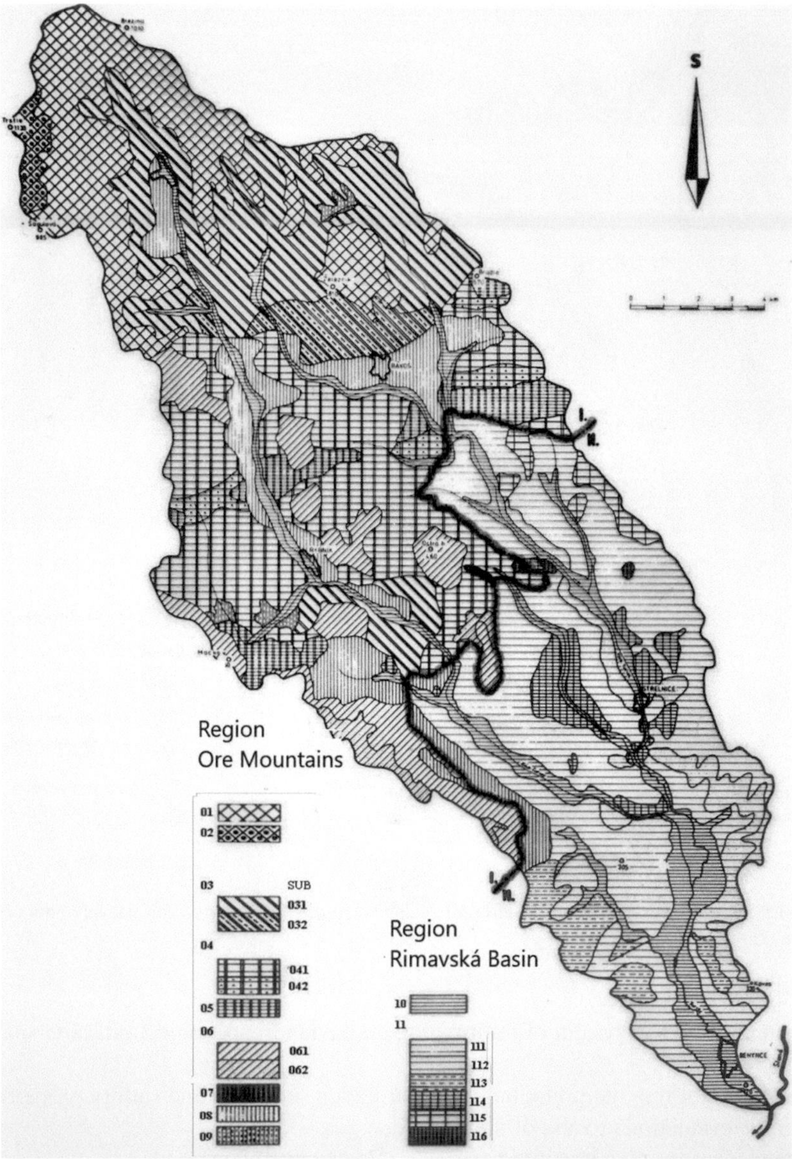

Fig. 4.4 Types and regions of landscape-ecological complexes in the Gemerské Turce catchment (original on the resolution level of map scale 1:50,000 *Legend* I. Region of Ore Mountains (**a**) Beach forest zone, types of landscape-ecological complexes TLEC: 01—of lower beach forest zone, 02—of higher beach forest zone (**b**) Oak forest zone, TLEC 03—on crystallinic rocks, subtypes 031—on rich substrates, 032—on poor substrates; 04—on mesozoic flysch rock, subtypes 041—on varied slates, 042—on marly slates; 05—on limestones; 06—on neovolcanities, subtypes 061—on volcanic slopes, 062—on volcanic plateau; 07—on radiolarites; 08—on proluvial sediments; 09—on slope debris. II. Region of the Rimavská Basin. TLEC: 10—on alluvial plains, 11—on hilly land; subtypes 111—on loess loams, 112—on gravels, 113—on marls, 114—on limestones, 115—on Mesozoic flysch rock, 116—on radiolarites

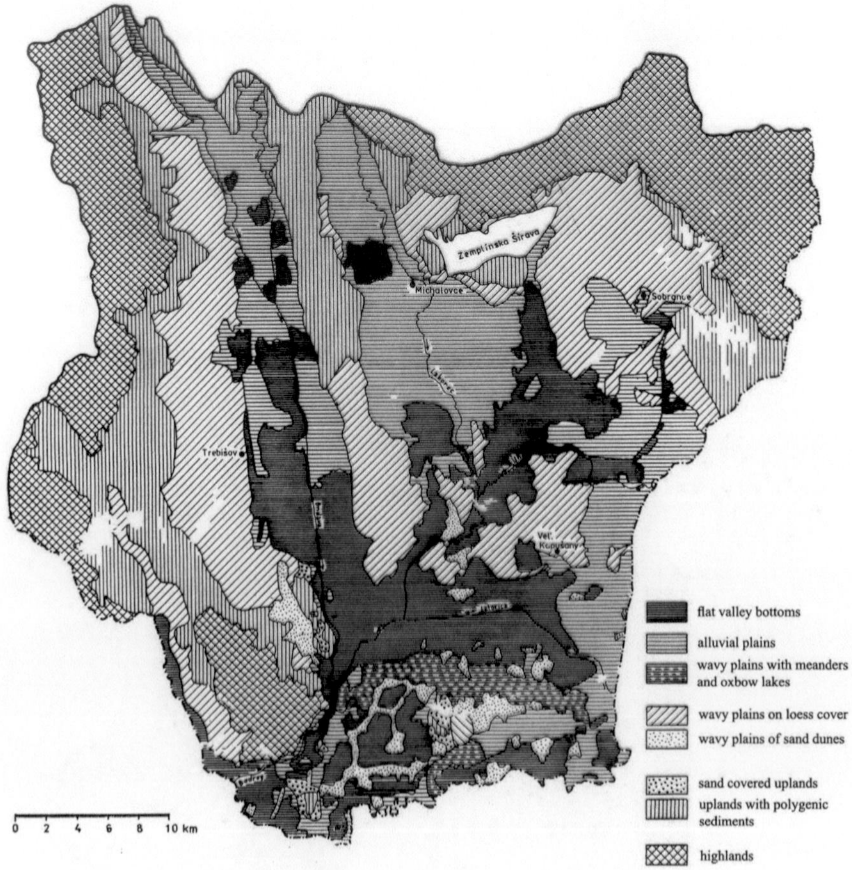

Fig. 4.5 Geosystems in the Eastern Slovak Lowland (originally on the resolution level of map scale 1:200,000)

- a comprehensive system of information on the landscape, object-attribute spatial database;
- the basis for the interpretations—the formation of the criteria (utility properties) for the evaluation process of the landscape;
- the object of evaluation with regards to selected human activities;
- the boundaries of the areas for the proposals of the optimal land use; therefore they are the subject of proposals.

As described above, the areas together with the characteristics of the types of geocomplexes creates the spatial databasis for different use. Since they are created on the background of the geosystem theory, for their creation, modern techniques of geographical information systems GIS are widely used. However, it is still not appropriate to rely completely on automatised process of the syntheses; first of all,

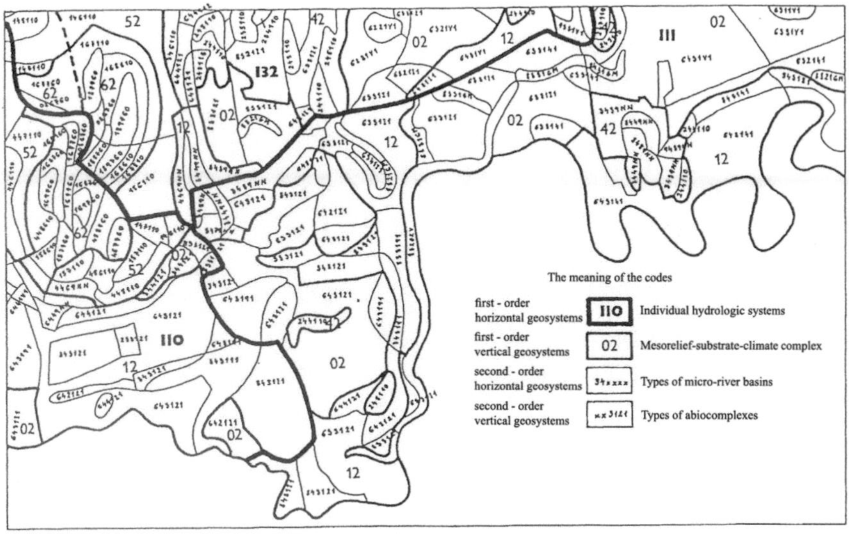

Fig. 4.6 Geosystems of various hierarchical orders and types in the Eastern Slovak Lowland (originally on the resolution level of the map scale 1:25,000)

because of disunity of cartographic base and of the topographic objects, as well as because of the inconsistency of the analytical materials produced by very different institutions. If we need complex and correct information and if we want to analyse the interaction of individual components, we have to standardise the analytical materials and we have to exclude the absurd combinations (Kocsis et al. 2016a, b; Konečný 2011). The basic elements of the topographic map and the basic elements of land use can be accurately projected into their uniform position, many times manually and with expertise as well as with the comprehensive understanding of the relationships between the components of the geosystem (Hreško et al. 2003).

Other aspect of the application of GIS is to be mentioned. The GIS technique allows to have a huge database on each point of a georeferencing base—actually landscape-ecological 'complexes' on each point—their indicator values can vary from point to point. Then, what is the reason for the creation of homogenous spatial units? First, because of the inconsistency of analytical materials, the "machine" can create also not realistic combinations of complexes, and, can create huge amount of very small areas of these non realistic complexes. Second, if our task is to propose

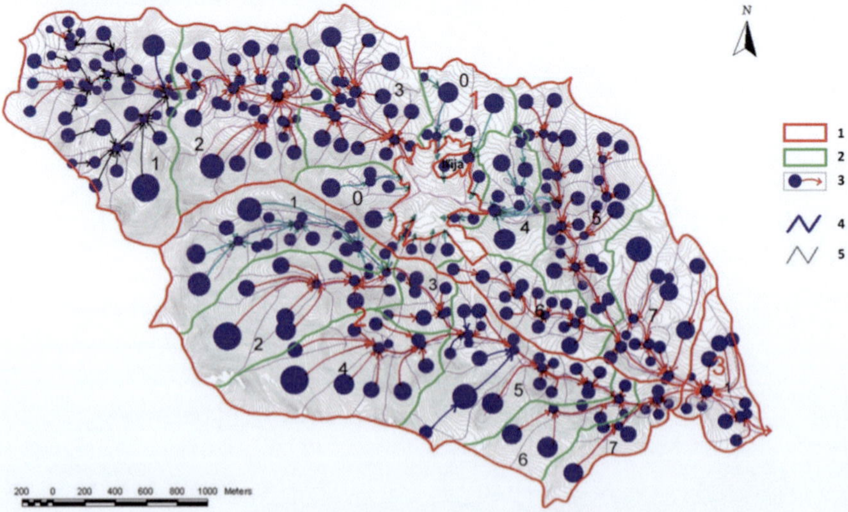

Fig. 4.7 Microcatchments of various order and their spatial interconnections (original resolution level on the map scale 1:10,000) *Legend* 1—microcatchment borders of second order; 2—microcatchment borders of third order; 3—borders and scheme of interconnections of microcatchments of fourth order; 4—watercourses; 5—contour lines

the optimum use of the territory, the areas of human activities—by simple words, the optimum land use—have their own demanded size, and are stable for years; therefore the creation of homogenous areas of landscape-ecological complexes as described above have still reasonable practical applications. Of course, the GIS techniques are nowadays inevitable tools for each research, as well as for each applied project.

Landscape-ecological syntheses in the LANDEP methodology represents:

- **the final step in the basic research, as well as**
- **the initial step in the applied procedures and planning practice.**

Landscape-ecological LEC complexes constitute:

- **the basis for the integrated information system**
- **the landscape-ecological basis for integrated land management**
- **the comprehensive spatial database for all steps of the planning processes.**

References

Armand DL (1964) Logichnost geograficheskikh klassifikaciy i skhem rayonirovaniya. In: Razvitie i preobrazovanie geograficheskoy sredi. Moskva

Bailey RG (2002) Ecoregion-based design for sustainability. Springer, New York, Nova Iorque

Csorba P (2008) A tájhatárok kijelölése és változása. Földrajzi Közlemények 132(2):220–226

Džatko M et al (1976) Charakteristika bonitovaných pôdno-ekologických jednotike SSR. MPVž, Bratislava

Hreško J, Mederly P, Petrovič F (2003) Landscape ecological research with support of GIS tools in preparation of landscape-ecological plan (model area of the Považská Bystrica city). Ekológia (Bratislava) 22(2):195–212

Hreško J, Bugár G, Petrovič F, Mederly P (2008) Endangering of abiotic complex as a basis for evaluation in the landscape-ecological plan of the Trenčín catastral area (Slovakia). Gospodarka surowcami mineralnymi 24(2):156–158

Kocsis K, Hevesi A (2003) A magyar-szlovák határvidék földrajza. Lilium Aurum, Dunajská Streda (Slovakia)

Kocsis K, Agárdi N, Koczó F, Mezei G, Zs Nemerkényi, Szabó R (2016a) MNA alaptérképétől a Web GIS-ig. In: Balázs B (ed) Theory meets practice in GIS (in Hungarian). Debreceni Egyetemi Kiadó, Debrecen

Kocsis K, Agárdi N, Koczó F, Mezei GA, Nemerkényi ZS, Szabó R (2016b) Generalizálás géppel, vagy kézzel? - módszertani összehasonlítás Magyarország Nemzeti Atlasza példáján. In: Pajtókné Tari I, Tóth A (eds) Magyar Földrajzi Napok: Absztraktkötet: VIII. Magyar Földrajzi Konferencia. Budapest: Magyar Földrajzi Társaság, p 103

Konečný M (2011) Dynamická geovizualizace v krízovém manažmentu. Masarykova univerzita, Brno

Łowicki D, Mizgajski A (2013) Typology of physical-geographical regions in Poland in line with land-cover structure and its changes in the years 1990–2006. Geographia Polonica, Warszawa 86(3):255–266

Mičian Ľ (1982) A system approach to landscape and the sciences realizing it to the system of geographical sciences. In: VIth international symposium on problems of landscape ecological research. October 1982, Piešťany, Institute of Experimental Biology and Ecology SAS, Bratislava

Miklós L (1978) Náčrt biologického plánu povodia Gemerských Turcov. VEDA, Bratislava, Questiones Geobiologicae 21:127

Miklós L, Kozová M, Ružička M et al (1986) Ekologický plán využívania Východoslovenskej nížiny v mierke 1:25 000. Ekologická optimalizácia využívania VSN. ÚEBE SAV Bratislava, Slovosivo. III. diel, pp 5–312

Miklós L, Miklisová D, Reháková Z (1986) Systematisation and automatisation of decision-making process in LANDEP method. Bratislava, Ecology (CSSR), 5(2):203–232

Špinerová A (2005) Charakteristika vybraných vlastností mikropovodí v povodí Ilijského potoka. Rigorózna práca. Univerzita Mateja Bela, Katedra krajinnej ekológie FPrV, Banská Bystrica

Špinerová A (2010) Krajinno-ekologické limity poľnohospodárskeho využitia Ilíjskeho potoka. VKÚ, a.s., Harmanec

Chapter 5
Landscape-ecological Interpretations in LANDEP

Abstract Interpretation in the LANDEP methodology is considered the process of formation of special-purpose characteristic of the landscape. The interpretations can have twofold nature accordingly to their further utilisation, as:

- as one of the steps of the LANDEP which results to creation of **criteria** and **arguments** for the further evaluation of the suitability of geocomplexes for the selected activities
- the character of final product of applied landscape-ecological research. In this case the interpretations may be considered as creation of special indicators of specific purpose-oriented properties, **utility values** of the landscape.

At present, interpretations represent the **mainstream** of the landscape-ecological research, regardless of whether they are called interpretations, landscape syntheses or evaluations. Landscape ecologists are working on diverse interpretations with endeavour to make use of their results in practice. The chapter intoduces several examples of interpretations developed for LANDEP in different periods on concrete model areas.The characteristics of technological-functional criteria—**localisation criteria**—derived from the **primary landscape structure** are used as criteria for location and technical performance of various activities. These criteria are closest to the technical standards. The chapter introduces example of such interpretations in concrete model territory, namely

- selected localisation criteria for agricultural land use,
- definition of the zones of slope dynamic—the conditions for the movement of water and material along the slope.

The characteristics of biological-ecological criteria—**selective criteria**—represent the evaluation of the **secondary/current landscape structure**, especially vegetation. These interpretations should provide arguments for the protection and conservation of nature and natural resources. The chapter introduces example of the

- expert interpretation of the degree of the ecological significance of vegetation
- ecological quality of spatial structure of land-cover
- spatial diversity of the landscape according to the degree of entropy.

L. Miklós and A. Špinerová, *Landscape-ecological Planning LANDEP*,
https://doi.org/10.1007/978-3-319-94021-2_5

The characteristics of socio-economic criteria—**realisation criteria**—are derived from the evaluation of **tertiary landscape structure**. They are based on the evaluation of the degree of current anthropogenic influence on the landscape, threat to the environment by human activities and legal protection of nature and natural resources. The chapter introduces example of the

- interpretations of socio-economic phenomena and their conflicts in the landscape
- interpretation of the protection degree of the landscape
- interpretation of the degree of load on the landscape.

Keywords Utility values · Localisation criteria · Selective criteria · Realisation criteria

Interpretation in the LANDEP methodology is considered the process of formation of special-purpose characteristics of the elements of the landscape-ecological complexes ***LEC***—interpreted indicators that will have twofold nature accordingly to their further utilisation, as:

- in the case the interpretation serves as one of the steps of the landscape-ecological planning, than the results have the sense of

criteria and **arguments**

for the further evaluation and determination of the suitability of an LEC for the selected activities within the planning process.

E.g. the land arability is a criterion for evaluation of the suitability of an abiocomplex for the proposal of its use as arable land, the erodibility of the BC is a warning criterion for planner not to propose intensive use, the ecological stability of a biotope is an argument of the value of a LEC for planners to keep it in original shape. This way is clearly practice oriented, as the part of planning process.

- in many cases the result of the interpretation have the character of final product of applied landscape-ecological research. In this case the process and the results of interpretations may be considered as creation of special

indicators of specific purpose-oriented properties

of the landscape.

In this case the erodibility, the arability, the ecological stability are as overall characteristics not necessarily bounded to further use. This way is more research oriented.

Character of interpretations

The interpreted special-purpose "properties" are not in fact physical properties of the complexes. They are mostly not measurable, not tangible, are more of a potential nature and can be called potential "properties" that enable us to determine the suitability of the LEC for the selected activities.

Both ways are very important for further development and improvement of both the applied research and the landscape ecological planning, too.

The outcome of the interpretations is actually a purpose-oriented re-thinking of either individual indicators, groups of indicators, or the whole LEC, and, their descriptions using the interpreted indicators that can be expressed in a number of ways, namely in words, on ordinal value scales up to calculated numerical values. The interpreted utility values/special purpose properties $\boldsymbol{j_n}$— when created within the methodics of LANDEP—are thus a function of the analytical properties $\boldsymbol{x_n}$ of the LEC types, formally described as

$$j_n = f\{LEC(x_1, x_2, \ldots x_n)\}.$$

It is to underline that the results of interpretations

- in terms of creation the **criteria and arguments** for the evaluation process in LANDEP as special-purpose properties of the landscape are to serve as criteria for decision on the suitability of ABC and LEC for human activities, as well as for enforcing landscape-ecological proposals or restrictions for the optimal utilization of the landscape. So, they are **functional part of LANDEP process**, as supporting step and background of evaluations and propositions
- in terms of creation the **utility values** are many times the final, **most desired products of the LANDEP** methodics, as well as generally desired result of landscape ecological research. Such results have also their own life, many times independent from LANDEP method, they are not involved further steps of LANDEP more. On other side they are used as arguments supporting the ecological ideas in very diverse studies.

So, generally, the process of interpretation of basic analytical and synthesized data on the landscape is of the character of applied landscape-ecological research. At present, methods of interpretation represent the **mainstream of the landscape-ecological research** regardless of whether they are called landscape-ecological interpretations, landscape syntheses or evaluations (Mander et al. 2005). Landscape ecologists are working on diverse interpretations with endeavour to make use of their results in practice.

The landscape ecologists assume that these results will be used in practice, but not all the results can be actually applied, not all are clearly defined or quantified. The problem is that these indicators are often immeasurable and hardly quantifiable, and in many cases are only the result of expert assessments or estimates, especially in case of **interpretations of the current landscape structure and vegetation**. Fuzzy characteristics and arguments might then be outweighed by technically and economically quantified criteria in a decision-making process. Another problematic aspect is the lack of definition on how to use and apply the results of the research in practice. Therefore, the scientific course to be pursued should consider enhancing the methods of landscape-ecological interpretations with competitive and cogent enough indicators.

The interpretations in terms of the creation of utility values of the landscape currently represent the mainstream of the applied landscape-ecological research.

Methods of interpretation

Methods of interpretation can be very diverse. The most commonly used are:

- **Expert assessments**—review of the analytical properties or the properties of the complexes—the process of wording of usually simple criteria for determining the suitability, e.g. the relation of the slope inclination as accessibility, soil depth and skeletality as arability; determining the state of ecological quality by indigenousness of vegetation etc. Simple expertise procedures developed into sophisticated ones, e.g. the methods of collective expertise, questionnaire expertise, yet sometimes they are merely based on the existing experience of the expert team (Kerényi and Csorba 1991).
- **Formalized assessments**—the essence of this methodology also lies in the expert review of the mutual relations between indicators while using formal tools, such as matrices, tables and specific rankings. In this way, it is possible to review the relations among more indicators in larger areas, e.g. determination of the tendency of water and material movement along the slope, evaluation of trophism, waterlogging, mechanical threat to soils and various ecological vegetation indicators (Hrnčiarová et al. 1996; Jurko 1986, 1990; Kalivodová and Kminiak 2003). They are also based on practical application of team´s knowledge and experience.
- **Calculations** of the values of special-purpose properties up to mathematical **modelling**, arising from the logical relations among indicators, which provides a base for formulas. Examples of such interpretations are, e.g. calculations of erosion threat (erodibility), evaluation of the potential amount of solar radiation, flood models, meteorological models etc. (e.g. Wischmeier and Smith 1978; Chlapečka 2004; Mander et al. 2005; Kočický et al 2008).

To this group of interpretation belong a huge amount of studies developing to so called landscape metrics (Turner and Gardner 1991; McGarigal 2002; Szabó et al. 2008, 2012).

Since the properties of geosystem indicators are of a very diverse nature, also the utility values/special purpose properties/interpreted criteria $\boldsymbol{j_n}$ are different. They are classified according to background analytical values they are based on, to the character of their utilisation in the evaluation and proposition. These groups are the following.

5.1 Technological-Functional Criteria—Localisation Criteria

They are used as criteria for evaluating the options of location and technical performance of various activities. These criteria are closest to the technical standards in the planning and projecting practice. They are mostly applied to abiotic complex elements, are relatively stable and most significantly affect the location of the required activities in the landscape; therefore, they can be called "technological-functional", or **location** criteria. (compare with e.g. Mariot 1983).

Such criteria include e.g. the potential of an area for being built-up, accessibility, cultivability, waterlogging, erodibility, supply of nutrients and water, risk of flooding, seismicity, carrying capacity of sub-soils and many others. The most important group of interpretations in this regard is the group of characteristics of erosion and accumulation threat, soil waterlogging and microclimatic conditions (Fu and Rich 2002; Kim 2006; Kočický et al. 2008). A popular direction is also the evaluation of the so called bearing capacity of the landscape (Hrnčiarová et al. 1996; Midriak 1999; Zaušková 2001).

The values of these indicators are often expressed as mathematical or physical quantities and are most extensively developed within the planning and projecting practice (Rich et al. 1994; De Roo et al. 2002; Németh et al. 2016).

Localisation criteria are derived from the elements of **the primary landscape structure**. They are therefore considered the steadiest and least changeable, which is why they should be uncompromisingly included in planning processes as arguments for decision making.

A specific group of special-purpose properties—assessment criteria—of the primary landscape structure is represented by the threat of natural hazards, sometimes referred to as natural disasters (Tremboš et al. 1994; Midriak 1997), such as volcanic zones, seismic zones, areas with a high risk of flooding, landslide zones. In our conditions they occured occasionally, so they should be assumed as potential (even negative one). The development of erosion and run-off models should be mentioned as very popular direction of interpretations based one the interpretations of abiotic complexes but modified also with the indicators of secondary landscape structure (Szeicz and Long 1969; Verrasztó 1979; Kočický et al. 2008).

Anyway, the potential occurrence of above mentioned processes at various intensity limits certain activities.

There are numerous methods of interpretations of the primary landscape structure that are constantly developing. For the LANDEP purposes, methods of **partially formalized procedures** of interpretations were developed in detail in the project of ecological optimisation of the Eastern Slovak Lowland (Miklós et al. 1986), but are still being developed today (Ružička et al. 1988; Izakovičová et al. 2010; Izakovičová and Moyzeová 2011; Špinerová 2015).

The following paragraphs will present two ways—examples—of the interpretations of localisation indicators, along with the projection of the results onto the maps of the Ilijský creek catchment area:

- a simple procedure of interpretations for the agricultural land use: interpretation of accessibility, waterlogging and cultivability;
- a more complicated procedure of interpretations of slope dynamics zones.

5.1.1 Example of Interpretations of Selected Localisation Criteria for Agricultural Land Use

In the Ilijský creek catchment area, as the result of landscape-ecological synthesis the types of landscape-ecological complexes entered to the process of interpretations. The LEC were defined as follows:

$$LEC\left[ABC\{x_1, x_2, x_3, x_4, x_5, x_6, x_7, x_8, x_9, x_{10}, x_{11}, x_{12}\}, \mathbf{CLS}(y_1)\right]$$

where the abiotic complexes ABC were defined by: x_1—quaternary geological bedrock—soil-forming substrate, x_2—hydrogeological types of sediments, weathering products and rocks, x_3—soil type (subtype), x_4—soil texture, x_5—soil depth, x_6—skeletality, x_7—relief inclination, x_8—relief orientation to the cardinal points, x_9—horizontal relief curvature, x_{10}—normal relief curvature, x_{11}—morphographic-position relief type, x_{12}—average altitude, the CLS by the current land use elements y_1.

The simple interpretation procedure was carried out to determine

(a) the localisation criteria of accessibility and soil cultivability.
They mainly result from the interpretation of the relief and soil properties.
(b) the selective and realization criteria: actual utility of LEC types.
This criterion results from the current landscape structure.

The procedure was as follows:

Accessibility—d

The interpreted criterion *accessibility* includes the conditions for machinery accessibility to areas, both in terms of movement and stability of machinery. In the existing database of LEC characteristics, accessibility—***d*** was characterized as a function of the analytical indicator x_1—relief inclination, and, the topographical position that can be identified based on element x_{11}—morphographic-position relief type. i.e.

$$\boldsymbol{d} = \mathrm{f}/x_7, x_{11}/.$$

The function is simple: the greater the inclination, the worse the accessibility. Given that the agricultural areas in the model area are situated roughly in the same location—in the middle or lower parts of gentle slopes, this indicator did not require any review or modification by the morphographic-positional types of relief.

Table 5.1 Simple verbal interpretation of accessibility

x_1—slope inclination	Verbal characterization	Degree of availability	Average % of tillage costs to costs with limited availability according to Table 5.4c
0°–1°	smooth	1.	67%
1°1′–3°		2.	70%
3° 1′–7°	good	3.	72%
7° 1′–12°	moderately restricted	4.	76%
12° 1′–17°	restricted	5.	93%
17°1′–25°	limited	6.	100%
over 25°1′	impossible	7.	Tillage is not expected

Numbers 1–7 indicate availability classes

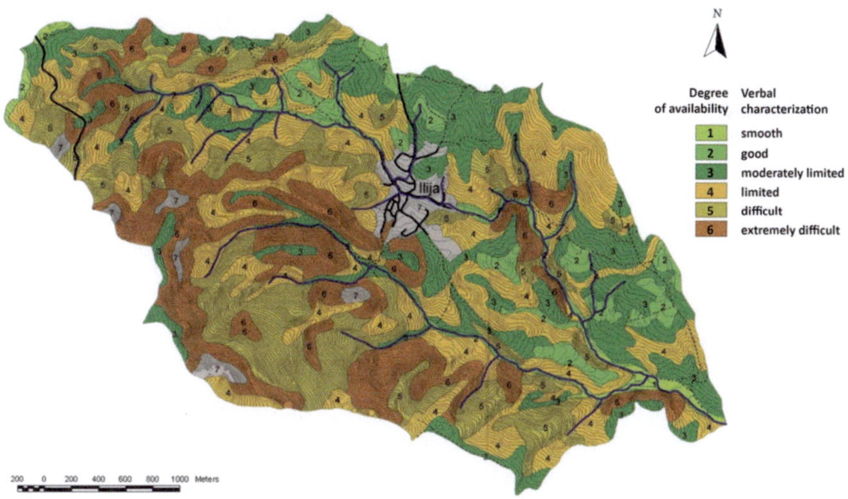

Fig. 5.1 Degree of accessibility (availability) for agricultural use

For this interpretation a partially formalised assessment method was used—a matrix.

Degrees of accessibility see on Table 5.1. Spatial interpretation of accessibility is on the map on Fig. 5.1.

Waterlogging by surface water—z

The interpreted criterion *waterlogging* includes also conditions for machinery accessibility to areas, both in terms of movement and stability of machinery, the time needed until the machines can enter the fields, as well as the condition for the presence of the surface water on the fields. In the existing database of LEC characteristics,

Table 5.2 Interpretation of waterlogging by surface water based on texture (permeability) and depression-inclination (surface run-off)

x_1—Slopes and depressed positions	x_4—Soil group				
	Light	Moderately heavy	Heavy	Very heavy	Clays
Saggy locations—2.5 m and more	2	7	9	9	9
Saggy—1.25–2.5 m	2	6	8	9	9
Saggy 0–1.25 m	1	5	7	8	9
0°–30°	0	4	6	7	8
30°–1°	0	3	5	6	7
1°–3°	0	2	4	5	6
3°–7°	0	1	3	4	5
7°–12°	0	1	2	3	4
12°–17°	0	1	1	2	3
17° and more	0	1	1	1	2

Numbers in the table mean classes of waterlogging with surface water 0—no waterlogging, 9—strongest waterlogging

waterlogging—z was characterized as a function of the analytical indicator x_4—soil texture (grain size), and x_7—relief inclination and depression, i.e.

$$z = \mathrm{f}/x_4, x_7/.$$

The function is simple, as well:

- the greater the inclination, the greater the surface run-off, the shorter the time of presence of water on the fields, and lower the level of waterlogging. In lowland areas, depression is an important indicator that points to the impossibility of surface water runoff. Obviously, the greater the depression, the longer the waterlogging.
- The heavier the soil is, the less permeable it is, and the greater is its susceptibility to waterlogging.

For this interpretation also partially formalised assessment method was used—a matrix. Degrees of waterlogging see on Table 5.2.

Cultivability—o

Cultivability includes the possibility of tillage and other agro-technical operations. In the existing database, cultivability $\boldsymbol{o}$ was characterized as a function of the analytical indicators, x_1—soil-forming substrate, x_4—soil texture, x_5—soil depth, x_6—soil skeletality, x_7—relief inclination, i.e.

$$\boldsymbol{o} = \mathrm{f}/x_1, x_4, x_5, x_6, x_7/.$$

The function means the following:

- the more stony and gravelly is interpretation of accessibility the soil-forming substrate contains, the more skeletal, less deep and less cultivable the soil is (Table 5.4a),
- the finer the texture (i.e. the heavier the soil), the worse the soil workability, the greater the waterlogging, the worse the seepage, and, consequently, the worse the cultivability.
- obviously, cultivability and workability is also affected by the terrain inclination. The greater the inclination, the worse the cultivability (Table 5.4b).

In this case, it is relatively clear, almost technically definable interpretation. Table 5.4a shows standardized rates per tillage hectare according to the soil depth and skeletality, Table 5.4b additional charges for tillage according to the slope inclination and texture (Miklós et al. 1986). The absolute numbers have obviously changed—they are still in CSK (Czechoslovak crowns), but the relative differences of numbers has remained the same, as it depends on the constant physical properties of abiocomplexes. Therefore, the ratio of these numbers can still be used, either in percentage, or as relative classes of cultivability. The combination of these values—as **complex cultivability**—is presented in Table 5.4c, where the percentages and cultivability classes are based on relatively objective physical properties of abiocomplexes.

These LEC indicators clearly specify the conditions not only for localization, or limitation of arable land, but also for activities where tillage is not a critical activity—e.g. the restrictive effect of soil-forming substrate and skeletality of the surface layer on mowing, the effect of soil depth on housing development.

For this interpretation the partially formalised assessment method was used—a matrix. Degrees of cultivability see on Table 5.3a–c. The spatial distribution on the model territory of Ilijský creek catchment area is shown on the Fig. 5.2.

5.1.2 Slope Dynamics Zones—The Conditions for the Movement of Water and Material Along the Slope

The dynamics of water and material movement down the slope (hereinafter referred to as "slope dynamics") in the process of the LANDEP is a purpose oriented property of great importance since influences several decisive processes, as the amount and dynamics of the run-off, the water erosion, retention capacity of the landscape. It also affects many other processes, e.g. accumulation processes, the flood situation, mechanical movement of material (see Miklós 1991, 1992; Miklós et al. 1986, 1997; Hrnčiarová 2001; Urbánek 2009).

The indicators affecting slope dynamics can be divided into three groups:

Table 5.3 **a** Interpretation of the cultivability according to the depth and skeletality

Soil depth	Soil skeletality			
	Without skeleton	Moderately skeletal	Very skeletal	Without soil
deep	1980	2280	2510	2900
moderately deep	2160	2530	2780	3250
shallow	2360	2750	3020	4000
without soil	–	–	–	4950

b Interpretation of the cultivability according to the texture and slope

Texture	Slope					
	Up to 1°	1°–3°	3°–7°	7°–12°	12°–17°	Over 17°
Light soil	363	373	303	444	600	700
Moderately heavy	400	428	441	489	640	700
Heavy	480	505	505	540	680	700
Very heavy	580	580	590	615	700	745
Clays	680	680	685	700	745	750

(continued)

Table 5.3 (continued)

c Interpretation of the complex cultivability according to the soil texture, skeletality, depth and slope angle

Soil texture →	Light	Light	Light	Moder. heavy	Moder. heavy	Moder. heavy	Light	Heavy	Moder. heavy	Heavy	Heavy	Very heavy	Very heavy
Slope →	do 1°	1°	3°	do 1°	1°	3°	7°	do 1°	7°	1°–3°	7°	do 3°	3°
Skeletality and depth ↓													
without or scarcely skeletal, deep	100	97.09	89.02	80.71	76.85	75.96	65.28	62.61	57.86	47.48	35.61	32.64	29.67
	1	1	1	2	2	3	3	4	4	5	6	7	7
without or scarcely skeletal, moderately deep	89.87	87.21	81.87	79.58	72.15	68.71	68.27	57.74	55.97	51.17	41.99	30.81	28.24
	2	2	2	3	3	4	4	5	5	5	6	7	8

(continued)

Table 5.3 (continued)

c Interpretation of the complex cultivability according to the soil texture, skeletality, depth and slope angle													
Soil texture →	Light	Light	Light	Moder. heavy	Moder. heavy	Moder. heavy	Light	Heavy	Moder. heavy	Heavy	Heavy	Very heavy	Very heavy
moderately skeletal, moder-ately deep	69.04	66.94	62.89	60.16	54.54	51.94	52.45	42.17	42.31	37.41	30.70	20.94	19.19
	4	4	4	4	5	5	5	6	6	7	7	8	9
moderately skeletal, shallow	56.69	55.01	51.64	48.65	44.11	41.99	43.07	33.00	34.21	29.25	24.0	15.87	13.83
	5	5	5	6	6	6	6	7	7	8	8	9	9
very skeletal, shallow	41.47	40.24	37.78	31.15	31.24	29.75	31.31	21.66	24.24	19.2	15.75	7.88	7.22
	6	6	7	7	7	8	7	8	8	9	9	10	10
skeletal, without soil	0.0	0.0	0.0	0.0	0.0	0.0	0.0	0.0	0.0	0.0	0.0	0.0	0.0
	0	0	0	0	0	0	0	0	0	0	0	0	0

(continued)

Table 5.3 (continued)

c Interpretation of the complex cultivability according to the soil texture, skeletality, depth and slope angle

Soil texture →	Light	Very heavy	Moder. heavy	Heavy	Clays	Clays	Light	Moder. heavy	Heavy	Very heavy	Clays	Clays	Clays
Slope →	12°	7°	12°	12°	≤ 3°	3°	17°	17°	17°	12–25°	7°	12°	17°
Skeletality and depth ↓													
without or scarcely skeletal, deep	25.22	17.8	5.93	5.93	4.45	0.0	0.0	0.0	0.0	0.0	0.0	0.0	0.0
	8	8	9	10	10	10	0	0	0	0	0	0	0
without or scarcely skeletal, moder-ately deep	26.67	21.82	15.92	5.25	4.91	3.68	0.0	0.0	0.0	0.0	0.0	0.0	0.0
	8	8	9	10	10	10	0	0	0	0	0	0	0

(continued)

Table 5.3 (continued)

c Interpretation of the complex cultivability according to the soil texture, skeletality, depth and slope angle

Soil texture →	Light	Very heavy	Moder. heavy	Heavy	Clays	Clays	Light	Moder. heavy	Heavy	Very heavy	Clays	Clays	Clays
moderately skeletal, moder-ately deep	20.49	14.83	12.83	3.84	2.8	2.1	0.0	0.0	0.0	0.0	0.0	0.0	0.0
	9	9	9	10	10	10	0	0	0	0	0	0	0
moderately skeletal, shallow	16.82	10.69	9.73	3.0	1.54	1.16	0.0	0.0	0.0	0.0	0.0	0.0	0.0
	9	9	10	10	10	10	0	0	0	0	0	0	0
very skeletal, shallow	12.31	5.58	6.89	1.97	0.0	0.0.	0.0	0.0	0.0	0.0	0.0	0.0	0.0
	9	10	10	10	0	0	0	0	0	0	0	0	0
skeletal, without soil	0.0	0.0	0.0	0.0	0.0	0.0	0.0	0.0	0.0	0.0	0.0	0.0	0.0
	0	0	0	0	0	0	0	0	0	0	0	0	0

Values in tables mean price of plowing in Kčs/hectare (Czechosloval crowns): lower price = better cultivability
Values in tables mean suitability percentage of cultivability towards the best (100%), as well as cultivability classes in a ten-step scale (1—optimal, 0—impossible cultivability)

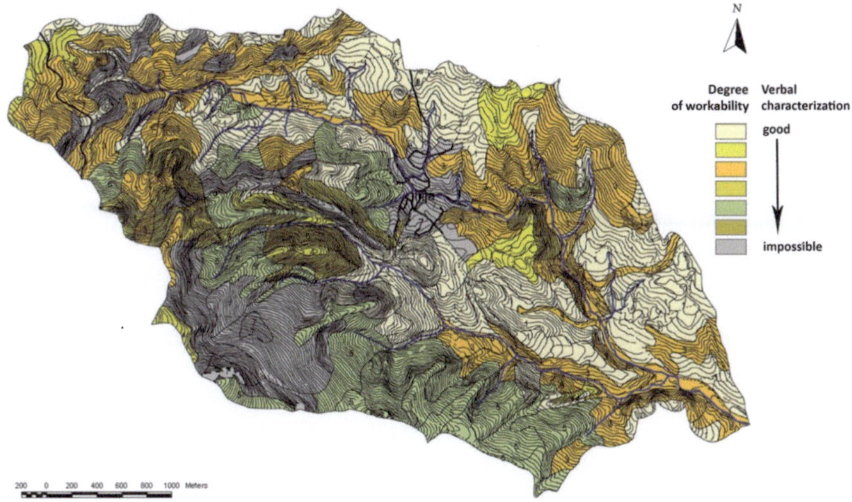

Fig. 5.2 Degree of cultivability for agricultural use

- stable abiotic indicators that do not change in the long run, especially the morphometric indictors of the georelief, geological bedrock, soil type and texture;
- variable indicators, particularly land use—although it can be said that some forms are stable from the perspective of a human's lifetime, e.g. forest area, built-up area, some may vary during the year—e.g. crop rotation on arable land;
- unstable indicators, changing indicators even in the short term, e.g. weather situation, soil moisture, soil water saturation.

Slope dynamics depend mainly on the morphometric relief indicators and their combinations that can be synthetically expressed in **morphotops**. In this case, morphotops are characterized within the types of abiotic complexes using indicators

$$\mathbf{MTOP}(\boldsymbol{x}_7, \boldsymbol{x}_8, \boldsymbol{x}_9, \boldsymbol{x}_{10}, \boldsymbol{x}_{11}, \boldsymbol{x}_{12}). \text{ (elements as in LEC)}$$

Let's discuss mainly the first two groups of indicators:

(a) **topographic position** of the studied morphotop (according to indicator $\boldsymbol{x}_{11}$—morphographic—position relief type).

It is a crucial factor affecting the basic **balance** of water and material movement down the slope that tells us whether to expect soil removal, both removal and supply, or prevailing accumulation in a particular morphotop, meaning that in positively isolated areas—on peaks, ridges, plateaus—there is only soil removal, in negatively isolated areas—in depressions, valley bottoms, vast plains—there is only accumulation, removal cannot be expected. In the other positions, both removal and supply in difference balance can be expected.

All other indicators make this basic scheme only more precise in terms of movement intensity, amount of material and course of processes, and based on these, it can be determined whether there is solely soil removal or supply, prevailing removal or supply, or they are in balance—in various combinations. The basic balance is easy to apply and from the point of view of sustainable development it is less important whether any process takes place within 2, 10, 50 or 100 years.

(b) **normal curvature** (according to indicator $\boldsymbol{x_{10}}$—normal curvature of relief)

It is an indicator of curvature of the gradient curves, which affects the **tendency** of the movement. I.e. we should expect acceleration on convex slopes, equable movement on uncurved slopes or deceleration of water and material movement on concave slopes—at constant values of the other indicators. Water and material movement objectively accelerates by gravity down the slope, but the normal curvature of morphotops can slow-down, steady or decelerate the movement tendency. E.g. if there is a morphotop with a convex slope, the movement will rapidly accelerate; if there is a morphotop with a straight slope, the acceleration will increase more slowly; and the acceleration on a concave slope may significantly slow down or even stop.

(c) **horizontal curvature** (according to indicator $\boldsymbol{x_9}$—horizontal curvature of relief)

It is an indicator of the curvature of contour lines that influences the direction of movement, i.e. dispersed movement on convex slopes, straight movement on straight slopes and concentrated movement on concave slopes. This indicator is relatively clear, well mappable. However, the natural curvature of the relief is often affected by anthropogenic lines—"harder" or "softer", that may significantly change the natural direction of movement.

In the actual landscape, in specific areas, we can observe a number of combinations of these indicators.

Besides these basic indicators, which characterize the stable, constant conditions of balance, tendency and direction of water and material movement down the slope, there is another important indicator of dynamics, namely:

(d) relief inclination (according to indicator $\boldsymbol{x_7}$—relief inclination).

Inclination along gradient curves determines the **intensity** (size, volume) and **time course** (speed) of water and material movement down the slope within the above-described processes.

The combination of the three pairs of morphometric indicators and interpreted characteristics of the their areas can be compared with catena of the slope (e.g. Urbánek 2009). We mark them characterize as

zones of slope dynamics.

The zones of slope dynamics are than significantly modified with fourth pair of indicators of intensity of the movement on slopes.

Indicator	***Interpretation***
x_{11}—Position ➤	***Balance: soil removal—accumulation***
x_{10}—Normal curvature ➤	***Tendency: acceleration—deceleration***
x_9—Horizontal curvature ➤	***Direction: dispersion—concentration***
↓	
Slope dynamics zones	
x_7—Inclination	***Intensity: speed, size, volume***

It must be stressed though that the slope inclination determines "only" the intensity and time course of the processes, not the nature of balance, tendency and direction of the movement.

Unlike the stable indicators for balance, tendency and direction of water and material movement down the slope, the movement intensity is affected by many other unstable indicators, such as intensity of rainfall, vegetation cover, water saturation of soils etc. That is where the importance of interpretation of slope dynamics zones lies, by which we can quite clearly determine the stable conditions of the slope dynamics while the other conditions are changing, and we can clearly identify the **relative differences in conditions** of adjacent areas that locally decide the optimal use of land.

5.1.3 Interpretation of the Slope Dynamics Zones in the Model Area

The interpretation and possible combinations of the indicators of position, normal curvature and horizontal curvature in the model area are shown in Table 5.4. The detailed analytical interpretation of morphotops from the point of view of the balance, tendency and direction of the downhill movement of water and material is shown on the Fig. 5.3. The coding and the colours on Table 5.4 serves as the explanation to Fig. 5.3.

By means of generalization of the detailed system interpretation of the analytical indicators obn Table 5.4 we can specify the slope dynamics zones as shown in Table 5.5.

The spatial distribution of the slope dynamics zones is shown on the map in Fig. 5.4. The colour-coding in the map key on Fig. 5.4 corresponds with the colour-coding in Table 5.5).

Modification of the slope dynamics zones by Inclination

Slope dynamics zones in the model area determined by balance, tendency and direction of water and material movement down the slope. They may be characterised in the next step according to the average inclination of LEC. As mentioned above, the inclination primarily affects the **intensity**, speed of movement or potential intensity of runoff, and the amount of material being transported. However, we must emphasise again that this is rather the characteristic of the time course of the processes, not

Table 5.4 Interpretations of position, normal curvature and horizontal curvature as balance, tendency and direction of water and material movement along the slope

Balance x_{11}—*topographic position*	Tendency of movement x_{10}—*normal curvature*	Direction of movement x_9—*horizontal curvature*		
		1. Dispersed convex slopes	2. Linear direction not-curved slope	3. Concentrated concave slopes
1.Material removal top, plateau	1. Acceleration convex slope	1.1.1 Accelerated floating- only removal		
2. Dominant removal crest	1.Acceleration convex slope	2.1.1 Accelerated dominant floating		
2. Dominant removal ridge	2. Even movement—straight slope	2.2.1 Accelerated floating	2.2.2 Dominant straight removal	
3. Balanced supply/removal upper slope	1. Acceleration convex slope	3.1.1 Accelerated balanced dispersed supply/removal	3.1.2 Accelerated balanced straight supply/removal	
3. Balanced supply/removal transport slope	2. Even movement—straight slope	3.2.1 Balanced dispersed supply/removal	3.2.2 Balanced straight supply/removal	3.2.3 Balanced concentrated supply/removal
3. Balanced supply/removal dell bottom	3. Deceleration—concave slope	3.3.1 Decelerated dispersed supply/removal	3.3.2 Decelerated straight supply/removal	3.3.3 Decelerated concentrated supply/removal
4. Dominant supply foothill	2. Even movement—straight slope			4.2.3 Dominant balanced supply
4. Dominant supply dellen bloc	3. Deceleration—concave slope			4.3.3 Dominant decelerated supply
5. Supply Foot, alluvium	3. Deceleration—concave slope			5.3.3 Decelerated concentrated supply
6. Only accumulation depression	3. Deceleration—concave slope			6.3.3 Only supply
7. Complex dynamics saddle, plateau			7.4.4 Complex motion dynamic	

the nature of balance, tendency and direction. Modifications of slope dynamics zones according to the movement intensity can also be interpreted as a degree of threat to these zones. The threat is characterized by the intensity of two degrees (zones of the average inclination up to 7° and over 7°), as well as by its nature (Table 5.6).

Modification of the slope dynamics zones by inclination

Slope dynamics zones in the model area determined by balance, tendency and direction of water and material movement down the slope. They may be characterised in the next step according to the average inclination of LEC. As mentioned above, the inclination primarily affects the **intensity**, speed of movement or potential intensity of runoff, and the amount of material being transported. However, we must emphasise again that this is rather the characteristic of the time course of the processes, not the nature of balance, tendency and direction.

Table 5.5 Slope dynamics zones

Zone code	Characteristic of slope dynamics zone	System characteristic	Code Table 5.5
1	Accelerated floating	Accelerated floating- only removal Accelerated dominant floating	111 211
2	Dominant straight removal	Dominant straight removal	222
3	Accelerated dispersed and straight transport	Accelerated balanced dispersed supply/removal Accelerated balanced straight supply/removal	311 312
4	Balanced transport	Balanced dispersed supply/removal Balanced straight supply/removal	321 322
5	Balanced and decelerated transport	Balanced concentrated supply/removal Decelerated dispersed supply/removal Decelerated straight supply/removal	323 331 332
6	Dominant concentrated movement and transport	Decelerated concentrated supply/removal Dominant balanced supply Dominant decelerated supply	333 423 433
7	Concentrated supply—water flood, material accumulation	Decelerated concentrated supply	533
8	Only supply—accumulation	Only supply	633
9	Complicated balance	Complicated motion dynamic	744

Table 5.6 Modification of slope dynamics zones according to the movement intensity and the most vulnerable zones of slope dynamics

		Intensity according to x_7—slope inclination great → small			
Zone code	Zone of slope dynamics	1 > 17°	2 7°–17°	3 3°–7°	4 0°–3°
1	Accelerated floating	1	1		
2	Dominant floated and straight material	1	1		
3	Accelerated dispersed and straight transport	2	2		
4	Balanced transport				
5	Balanced and decelerated transport				
6	Dominant concentrated movement and transport—potential for gullies	3	3		
7	Concentrated supply—water flood, material accumulation				4
8	Only supply—accumulation				
9	Complex balance				

1–4 and colours—the most endangered zones of slope dynamics. Other zones. Grey coloured—other existing combinations in model territory we consider un-endangered

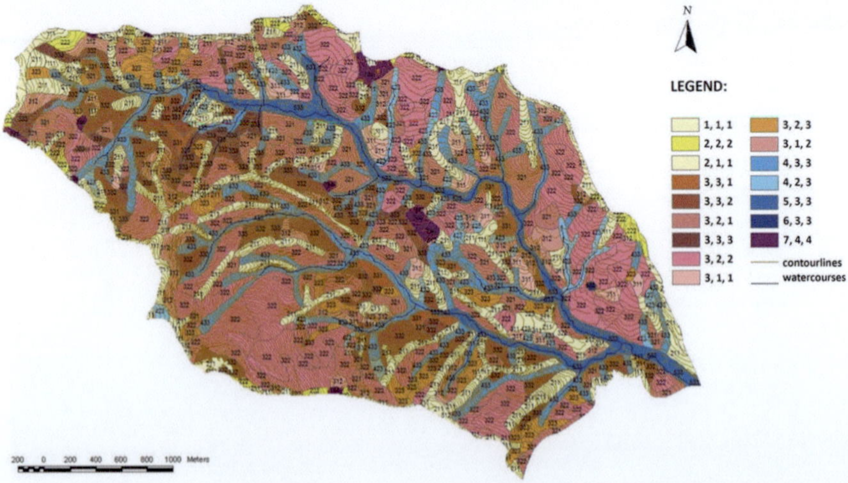

Fig. 5.3 Interpretation of morphotops from the point of view of the balance, tendency and direction of the downhill movement of water and material *Legend to Fig.* 5.3*: 1st code—balance according to position; 2nd code—tendency according to normal curvature; 3rd code—direction according to the horizontal curvature. Explanation see Table* 5.4

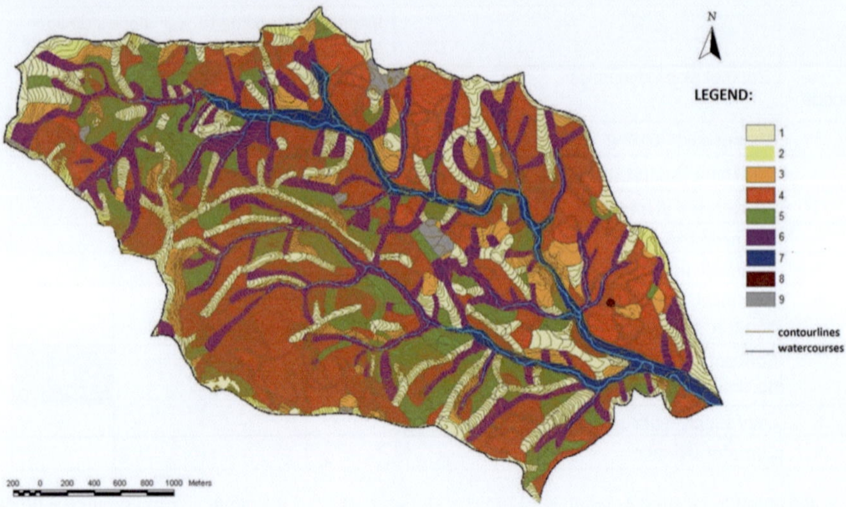

Fig. 5.4 Zones of slope dynamics Explanation of colours see in Table 5.5 and in the text

Modifications of slope dynamics zones according to the movement intensity can also be interpreted as a degree of threat to these zones. The threat is characterized by the intensity of two degrees (zones of the average inclination up to 7° and over 7°), as well as by its nature.

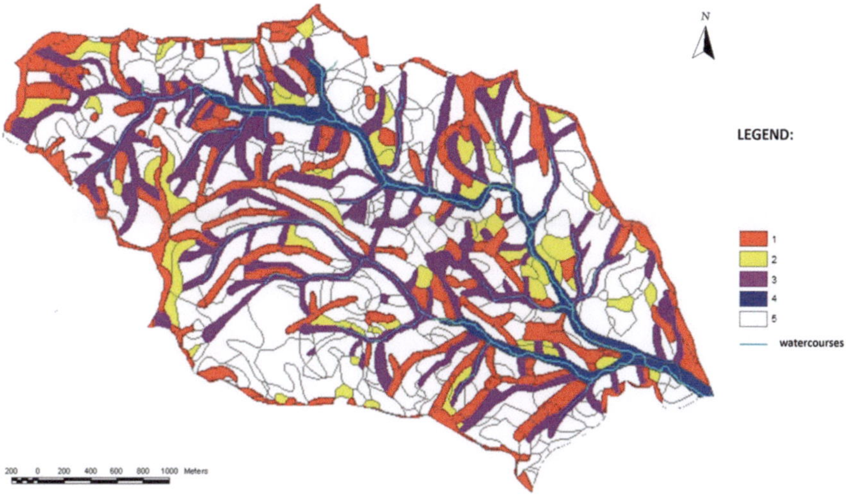

Fig. 5.5 Most endangered slope dynamics zones. Explanation of colours see in Table 5.6 and in the text

The zones according to the modification may be defined by following characteristics of slope dynamic (the numbers correspond to the key of the legend of the Fig. 5.5):

1. Accelerated floating and material removal, erosion (zones 1 and 2, combinations 111, 211, 221, 222).
2. Accelerated dispersed and straight transport, sheet water erosion (zone 3, combinations 311, 312).
3. Concentrated water and material movement—potential formation of gullies (zone 6, combinations 333, 423, 433).
4. Concentrated supply—excessive accumulation of material and flooding in the case of flash floods (zone 7, combination 533).
5. The other zones, without specific processes or threats.

Modification of the zones of slope dynamics by elements of CLS (land cover)

Zones of slope dynamics are of stable character and will continue to function as described above in case of any cover. However, their course will have varying intensity, in case of arable land the intensity can vary throughout the year.

The elements of CLS may affect the surface runoff mainly by the following ways:

- retaining surface runoff by infiltration. This function can be also artificially promoted by planting infiltration belts;
- changing the direction of surface runoff. They include especially various CLS elements, which are of either a “soft” character e.g. the edge of a forest or lines of shrubs of hard character e.g. roads and other technical structures commonly accompanied by drainage ditches.

Table 5.7 Real threats to slope dynamics zones from current landscape structure CLS

		Modification of slope dynamics zones according to CLS		
Intensity of slope dynamics zones threat (according to Table 5.7)		1. Exponentiate dynamics—*arable land, built-up areas, residential areas*	2. Invariable dynamics—*heterogenous agricultural areas, permanent cultures*	3. Decelerate dynamics—*meadows, pastures, scrubs, forests*
1	**Accelerated floating and material remova, erosion (zones 1 and 2, comb. 111, 211, 221, 222).**	1.1	1.2.	1.3
2	**Accelerated dispersed and straight transport, flat erosion (zone 3, comb. 311, 312)**	2.1.	2.2.	2.3
3	**Concentrated water and material movement – potential formation of gullies (zone 6, comb. 333, 423, 433).**	3.1.	3.2	3.3
4	**Concentrated supply – excessive accumulation and flooding by water floods (zone 7, comb. 533)**	4.1	4.2	4.3
5	**Other zones, without specific processes and threat**	5.1	5.2	5.3

According to modification of the zones of slope dynamics, they can be divided into three categories, namely:

- zones with enhanced dynamics: zones with arable land, built-up areas, residential areas where water movement is not hampered by vegetation,
- zones with unmodified dynamics: zone with permanent grasslands and heterogeneous agricultural areas with crop rotation,
- zones with impeded dynamics: forested areas, with greenery and grasslands, movement of water and thus transport of material is assumed to be hampered by vegetation cover.

We reiterate that this is only a modification which affects the course of processes, but not a change of the zones of slope dynamics. This modification represents the characteristics of real threat to slope dynamics zone (compared to potential threat based just on morphometric conditions). It is presented in Table 5.7.

The real threat to slope dynamics zones can be verbally interpreted as follows (the numbering correspond to the Table 5.7. This verbal interpretation serves as the explanation to the Fig. 5.6):

1.1. Significant threat from accelerated floating and removal of material (zones 111, 211, 221, 222 with arable land).
1.2. Moderate threat from accelerated floating and removal of material (zones 111, 211, 221, 222 with heterogeneous agricultural areas and permanent crops).
2.1. Significant threats from accelerated dispersing and unidirectional transport (zones 311, 312 with arable land).
2.2. Moderate threat from accelerated dispersing and unidirectional transport (zones 311, 312 with heterogeneous agricultural areas and permanent crops).

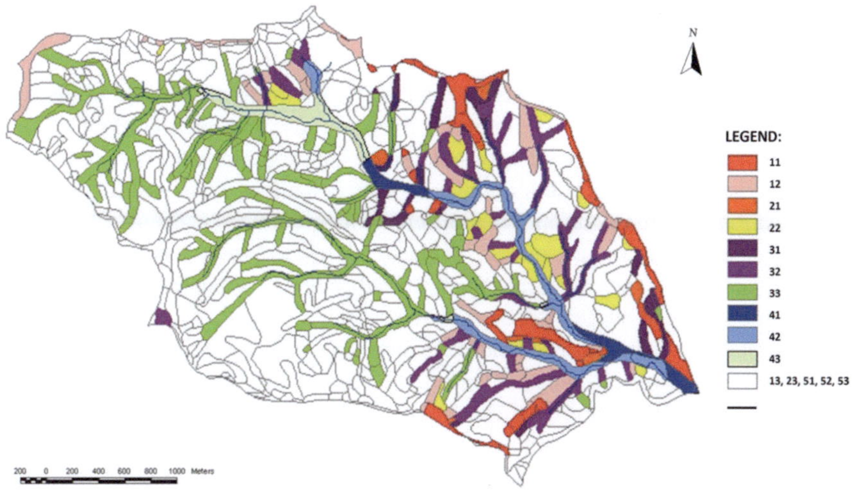

Fig. 5.6 Real threat of slope dynamics zones according to the current land use Explanation of the colours see Table 5.7 and in the text

3.1. Significant threat from concentrated movement of water and material—possible formation of gullies (zones 333, 423, 433 on agricultural soils).
3.2. Moderate threat from concentrated movement of water and material—precursor of formation of gullies in case of sustained exposure of soil (zones 333, 423, 433 with heterogeneous agricultural areas and permanent crops).
3.3. Slight threat from concentrated movement of water and material—possible formation of gullies in case of poor forest management (areas 333, 423, 433).
4.1. Threat from flash floods and possible accumulation of material (zone 533 settlements, arable land).
4.2. Slight threat from flash floods and accumulation of material in other agricultural areas, in permanent crops (zone 533 in other agricultural soils).
5. All other zones without specific processes and threats.

The spatial distribution of threats and real threats to slope dynamics zones is shown in Figs. 5.5 and 5.6. The colours in the map key correspond with the colours in Tables 5.6 and 5.7.

Such a character of the results of the interpretations can serve as a criterion for determining the suitability of the utilisation of areas and for proposals for measures and optimisation their implementation to mitigate existing problems.

5.2 Biological-Ecological Criteria—Selective Criteria

These represent the criteria which issue from the evaluation of the **current landscape structure CLS**, especially vegetation, for the purposes of protection of nature and natural resources. These interpretations should provide arguments for their protection and conservation. They are less definite and often without the possibility of their quantification thus they are open for different interpretation. Currently, there is a general consensus on the acceptance of these criteria to preserve biodiversity and promote sustainable development. We use them mainly to determine the special-purpose properties of biotic landscape elements.

Interpretation of the current landscape structure and vegetation are among the core areas of landscape ecology. There is a large number of similar methods, the LANDEP methodology, however, can adopt all the methods which allow the evaluation of the suitability of current land use and propose its optimal utilisation. These interpretations are particularly important for the design and projecting of territorial systems of ecological stability where they function as an argument to determine, which CLS elements are suitable to perform some of the required functions of **biocentres, biocorridors and interactive elements**, but also **ecostabilising measures**.

In principle, the interpretative methods of CLS and vegetation focus on the following categories of characteristics:

- structural characteristics of the vegetation: e.g. species richness, life forms, life strategies.
- state characteristics of vegetation: e.g. stability, degree of hemeroby, vulnerability, carrying capacity, balance.
- ecosozological characteristics of vegetation: regional rarity.
- production properties of vegetation: fodder value, melliferous potential.
- significance of vegetation: namely landscape ecological significance, nature-conservation significance, species richness, naturalness, regional rarity of species, biotic significance (habitat significance), gene pool values, degree of threat, degree of synanthropy.
- ecological-stabilising and landscape-ameliorative properties of vegetation: degree of ecological stability of CLS elements, ecological carrying capacity.
- ecological quality of spatial landscape structure.
- non-productive functions of biotic elements: socio-economic values, anti-erosion, climate, water, hygiene, aesthetic importance (Papánek 1974,1978; Jurko 1990; Zachar et al. 1989; Štefunková and Cebecauer 2006, a i.).

This group of criteria argues for **maintaining of a specific element** in selected sites and performance of activities on a selective basis, hence the name—**selective** indicators of landscape properties.

These criteria are difficult to articulate into a strictly defined legal form. Even legally and economically defined values—e.g. the ideological value of a protected animal—are just a convention, and not physically or mathematically derived values (Maglocký 1993). In planning procedures, the final values acquired from any complex and formalised evaluation process of assessment using multi-range scales are usually awarded a binary value: "suitable-unsuitable", to be preserved—not to be preserved".

Even the criteria already incorporated in the legal regulations e.g. "ecological stability" in the Law on forests, land consolidations, land-use planning are not easy to apply in practice.

Important principle of decision-making in relation to selective criteria

Selective criteria are derived from elements of the **secondary landscape structure**. These are changeable by exerting a certain amount of energy and therefore are the main object mankind tries to change. In this sense, their functional values and limits are relatively unstable (e.g. while a small forest exists, it can have a high ecological and functional value for the society, as soon as it is cut down, this value ceases to exist). Therefore, it is vital to recognise the selective importance of this indicator as an **argument** and **criteria** for various human activities in the landscape (Odum 1977; Kontriš 1978; Šteffek et al. 1992).

The most significant methods of interpretations of CLS and vegetation in landscape-ecological planning are those, which can be spatially projected and thus meet the requirement for the **planning** documents.

In this regard, the most important interpretation groups are:

- ecosozoological characteristics of vegetation: regional rarity, vulnerability.
- significance of vegetation: namely landscape ecological significance, nature-conservation significance of vegetation, species richness, naturalness, vulnerability and regional rarity of species, biotic significance of vegetation (habitat significance), *gene pool values, rarity, degree of threat, degree of synanthropy.*
- ecological-stabilising and landscape-ameliorative properties of vegetation: degree of ecological stability of CLS elements, ecological carrying capacity.
- ecological quality of spatial landscape structure.
- ecologically significant landscape elements.

Detailed description of the interpretation methods for the purposes of planning practice can be found in e.g. Midriak (1999), Izakovičová et al. (2000), Kalivodová and Kminiak (2003), Kozová and Bedrna (2003), Moyzeová and Grotkovská (2006), Špulerová (2010), Diviaková (2010, 2011), Špulerová et al. (2013), Pauditšová (2010), Miklós et al. (2011), and many others. At this point we briefly present only some of the examples of frequently used characteristics in the LANDEP methodology, namely

- methods of special-purpose interpretations of vegetation significance and CLS,
- methods of the evaluation of the ecological quality and diversity of spatial landscape structure.

5.2.1 Special-Purpose Expert Interpretation of the Significance of Vegetation and Biotic Elements of CLS

Interpretations of vegetation offer a wide range of special-purpose classifications. These methods of expert evaluation draw on the experience of researchers gained from practical field research. Of course, many authors have attempted to formulate the interpretation process, the expert character of the evaluation has, however, endured (e.g. Papánek 1974, 1978; Zachar et al. 1989; Jurko 1986, 1990; Míchal 1992; Löw et al. 1995; Midriak 1999; Halada 1998; Halada et al. 1995; Hreško 1996). In terms of the LANDEP methodology, especially the works on the methodology of carrying capacity of the landscape caused a substantial advancement in these methodologies (Hrnčiarová et al. 1997; Hrnčiarová 1999a, b).

The method of expert evaluation is relatively simple. The main principle is a clear content and spatial definition of the units assessed—vegetation units, habitats, CLS elements as well as their logical categorisation into a scale based on a selected leading factor and subsequent assignment of values of special-purpose characteristics to these unit.

As an example of such expert evaluation we present the interpretation charts of ecological significance of CLS elements, vulnerability of vegetation and CSL elements for the model territory of the Eastern Carpathians (Hrnčiarová et al. 1997, the expert team leader was RNDr. Helena Ružičková, CSc.) (Tables 5.8 and 5.9).

5.2.2 Ecological Quality of Spatial Landscape Structure

Determination of ecological quality of spatial landscape structure is of substantial **argumentative significance** in landscape-planning procedures. It can quantitatively determine the need to designate new ecologically more stable landscape elements in particular territorial units—districts, villages, microcatchment areas.

Also in this case the types of landscape-ecological complexes LEC

$$\boldsymbol{LEC}\big[\boldsymbol{ABC}\{\boldsymbol{x}_1, \boldsymbol{x}_2, \ldots, \boldsymbol{x}_{12}\}, \mathbf{CLS}\{y_1\}\big]$$

act as the output for the interpretations which rely on re-evaluations of the properties of the elements of **CLS** $\{y_1\}$. As most methods employed to determine a degree of ecological stability is underlain by a consideration that the higher ecological quality a particular element of current landscape structure (CLS) has, according to biological, ecological, but also other utilitarian functions, their **positive spatial impact** becomes more prominent e.g. have a higher water-retention, soil erosion and filter capacity, higher safety, hygienic and aesthetic functions. A similar procedure was adopted

Table 5.8 Ecological significance of CLS elements: model territory of the Eastern Carpathians

Elements of the land use	Ecological significance
Herbaceous—grass vegetation	
Uncultivated meadows and pastures, fallows (05)	1–2
Peat and muddy meadows (06)	1–2
Lush meadows and pastures (03)	2
Dry and semi-dry grasslands (09)	2–3
Meso- and oligotrophic *grasslands* (10)	2–3
Poloniny—Eastern Carpathians alpine meadows (11)	2–1
Reclaimed meadows and intensively grazed pastures (12)	3
Complex of herbaceous—grass and woody vegetation	
Complex of shrub vegetation (in groups and lines) with understory level	2
Meadows and pastures with advanced successional stages over 50% volunteer plants of the area (spruce, birch, pine and others)	2–3
Woodland areas (forests and shrubs on forest and agricultural lands)	
Alder forests on alluviums of flats and flows (28, 40)	1
Oak-Hornbeam forests (24)	2
Calcifuge oak and beech forests (25, 38)	2
Beech forests (27, 30, 34)	2
Lime and maple forests on slopes, screes and ravines (29, 37)	2
Spruce forests (36)	2
Agricultural elements	
Arable land with big blocks (46)	5
Arable land with small blocks (47)	3
Narrow fields, meadows, non- forest vegetation with varying % of representation	1
Gardens, plantations, orchards (without objects) (49)	2
Abandoned agricultural areas (53)	3
Water flows and water areas	
Water flows[7] (54a)	1
Lakes (56)	1
Artificial water areas (dams) (57)	1
Industrial, energy industry and transport elements	
Areas of plants (industrial objects, stocks, boiler rooms) i.) (61)	3
Overhead power lines (64)	3
Roads (67)	3

(continued)

Table 5.8 (continued)

Elements of the land use	Ecological significance
Elements of field-, forest- and water management	
Areas of agricultural holdings (71)	4
Forest nurseries (73)	1
Clearcuts	5
Residential, recreational, resting, sporting and cultural-historical features	
Residential areas—prevalence of family houses and gardens over 50% of the area (79)	2
Ski lifts (88)	3
Areas of ethnographic celebrations (97)	3
Memorials, tumulus, lookouts (0A)	2
Sacred buildings (churches, wayside shrine) (0B)	1

Leader of the expert team RNDr. Helena Ružičková, CSc
Ecological significance of the CLS element:
1—very significant, 2—significant, 3—moderately significant, 4—less significant, 5—not significant

already for the project of the Environmental Master Plan of the Slovak Socialist Republic in 1986, where the knowledge of the ecological quality of CLS elements, acquired through many years of expert experience, was formulated as a coefficient (Jurko 1990, Miklós 1986), which modifies the areas of individual CLS elements as indicated by the formula:

$$\boldsymbol{K}_{\mathbf{es}} = \left(\Sigma \boldsymbol{p}_i.\ \boldsymbol{k}_{\mathrm{SKSi}}\right).\ \boldsymbol{P}^{-1},$$

where

$\boldsymbol{K}_{\mathbf{es}}$ coefficient of the ecological quality of the territory
$\boldsymbol{p}_i$ area of the $\boldsymbol{i}$-th CLS element,
$\boldsymbol{k}_{\mathbf{SKSi}}$ coefficient of ecological stability of the $\boldsymbol{i}$-th CLS element
$\boldsymbol{P}^{-1}$ total area of the examined territory

The final value of the ecological quality of the examined territory is represented by a weighting of an area by coefficient of ecological stability what results to the estimation of rate of the **ecological-quality area** from the total area of the examined territory. This rate is of course, smaller than the total area of the territory. The interpretation, thus, assumes a smaller area for ecological services than the total area, which is expressed by weighting ratio. Similarly, all other modifications of the ecologically quality area assume its reduction in this sense (see below).

An example of determination of $\boldsymbol{K}_{\mathbf{es}}$ for respective mapping units of CLS is shown in Table 5.10 and Fig. 5.7.

Table 5.9 Vulnerability of the real vegetation types by selected disturbing factors: model territory of the Eastern Carpathians (Leader of the expert team RNDr. H. Ružičková, CSc.)

	Vulnerability by selected disturbing factors				
The real vegetation types and elements of the land use	Mechanical disruption of the soil surface	Change of underground water level	Chemical load on the environment (eutrophication)	Absence or change of traditional land use	Removal or destruction of the vegetation
Vegetation of stagnant waters and marshes	**0**	2–3	2	**0**	2
Complex of non-forest woody vegetation, wet and muddy meadows	2	2–3	2–3	1	2–3
Uncultivated meadows and pastures, meadow fallows	2	1–3	2	2	1–3
peat and muddy meadows	**0**	2–3	3	201	3
Wet meadows and pastures	2	2–3	2–3	2–3	2–3
Lush meadows and pastures	2	1	2	2	2
Dry and semi-dry grasslands	2–3	**0**	2	1–2	2–3
Meso- and oligotrophic grasslands	1–2		3	1–2	2–3
Poloniny: Eastern Carpathians alpine meadows	3	**0**	2–3	2–3	2–3
Reclaimed meadows and intensively grazed pastures	1–2	1	1	2	1
Vegetation of rock surface and scree slopes	3		**0**	1	2–3
Subalpine herbaceous and shrubs	3		3	**0**	2–3

(continued)

Table 5.9 (continued)

	Vulnerability by selected disturbing factors				
The real vegetation types and elements of the land use	Mechanical disruption of the soil surface	Change of underground water level	Chemical load on the environment (eutrophication)	Absence or change of traditional land use	Removal or destruction of the vegetation
Alpine herbaceous and shrubby vegetation	3		3	**0**	3
Alder forests on alluviums of flats and flows	3	2	2–3	**0**	3
Birch—Alder forests	3	3	3	**0**	3
Floodplain forests—soft, transient, hard	2	2–3	2	**0**	2–3
Oak-hornbeam forests	1	**0**	1	**0**	1
Calcifuge oak and beech forests	1		2	**0**	2
Beech forests	1		2	**0**	1
Fir forests	2	1	2	**0**	2
Lime—maple forests on screes and ravines	2	**0**	3	**0**	2–3
Spruce forests	2	**0**	2	**0**	1–2
Peat and waterlogged spruce forests	3	2	3	**0**	3
Knee timber forests	3	**0**	3		3
Poplar monocultures	1	2	2	**0**	1
Other deciduous monocultures	1	**0**	2		1
Coniferous monocultures	1	**0**	2	**0**	1

Vulnerability verbal characteristics:
0—vegetation does not respond to a given factor, 1—vegetation less vulnerable to a given factor, 2—medium-vulnerable vegetation, 3—very vulnerable vegetation

Table 5.10 Coeficient of ecological quality and run-off coefficient of CLS elements (Ilijský creek catchment)

i	Element of CLS	Area in m^2	% of area	k_{CLSi}	k_{odi}
1.	Deciduous forest	2,260,875	14.5	1	0,03
2.	Mixed broad-leaved deciduous forest	1,010,146	25.4	0,95	0,03
3.	Mixed forest with balanced stand composition	3,966,003	6.5	0,9	0,04
4.	Mixed forests with prevalence of coniferous trees	215,643	1.4	0,85	0,05
5.	Coniferous forest	244,187	1.6	0.75	0.06
6.	Continuous scrub vegetation	302,152	1.9	0.7	0.1
7.	Water *s*	3540	0.02	0.6	1
8.	Mosaic of grasslands and scrub vegetation	2,101,958	13.5	0.7	0.13
9.	Meadows and pastures	1,091,530	7.0	0.65	0.15
10.	Orchards	334,157	2.1	0.5	0.13
11.	Arable land	3,646,209	23.4	0.25	0.25
12.	Greenery in residential areas	97,180	0.6	0.4	0.13
13.	Sports areas	1236	0.01	0.25	0.2
14.	Buildings and other technical elements	52,384	0.3	0	1
15.	Other areas in settlements, courtyards	157,479	1,0	0,1	0,7
16.	Landfills	4785	0,03	0,1	0,5
17.	Reinforced structures (roads, parking places)	109,853	0,7	0	1
18.	Rocks, rock slopes, terrain notch	1738	0,04	0,2	0,8

k_{CLSi}—coefficient of ecological quality ***i***-th element of CLS
k_{odi}—runoff and retention coefficient for erosion models

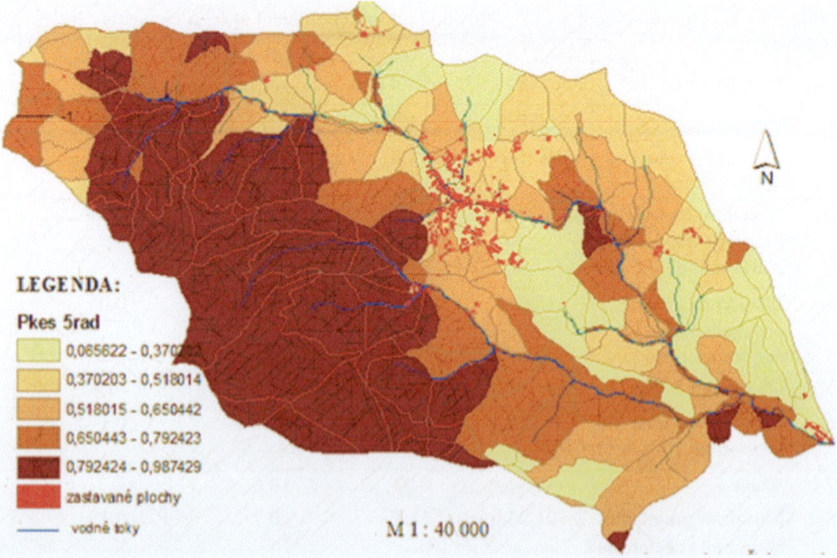

Fig. 5.7 Coefficient of ecological quality in the microcatchments of the Ilijský creek catchment, *The bigger number (the rate of the ecologically quality area of microcatchments), the darker colour, the higher quality. The red colour marks the built-up areas*

Modification of the assessment of ecological quality of the territories of municipalities

The above described fundamental procedure of the interpretation can be further modified and interpreted in various ways and for different territorial units. The basic starting point in this case is identical with the above discussed example. The most often used procedure is the characteristic of the ecological quality of the cadaster territories of settlements. It is possible

- to calculate the values of ecological quality for different time periods, which allows to compare the change of ecological quality in time (Fig. 5.8)
- to modify the area of ecological quality with geomorphological units, with climate comfort and environmental state, protected areas (Fig. 5.9)
- calculate these values per capita, and many other.

Few examples are demonstrated on the example of cadastral areas of the catchment area of the river Ipel' (Miklós et al. 2011).

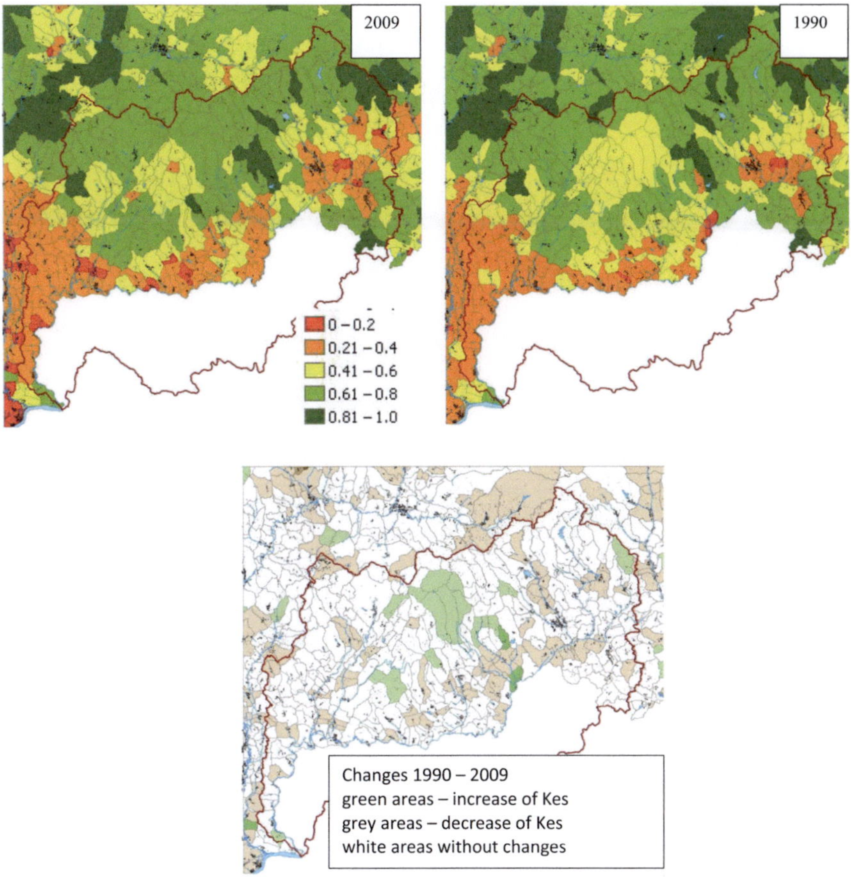

Fig. 5.8 Coefficient of ecological quality of cadastral territories and their changes in the Ipel' river basin

5.2.3 Spatial Diversity of the Landscape According to the Degree of Entropy

An often used and well interpretable indicator of spatial landscape structure is entropy that is defined as a degree of disorderliness. In the spatial context, entropy can be interpreted as follows—the higher a degree of entropy, the higher disorderliness and thus also the diversity of an examined territory (Hrnčiarová 2009). Here, we utilise the "classical" Shannon's entropy equation. The basic way of its application in landscape planning is the estimation of probability of occurrence of $\boldsymbol{p_i}$ of individual areas of landscape-ecological complexes (LEC) which enter the basic formula as $\boldsymbol{\Delta p_i}$ within the model territory $\boldsymbol{P}$. The basic equity reads as

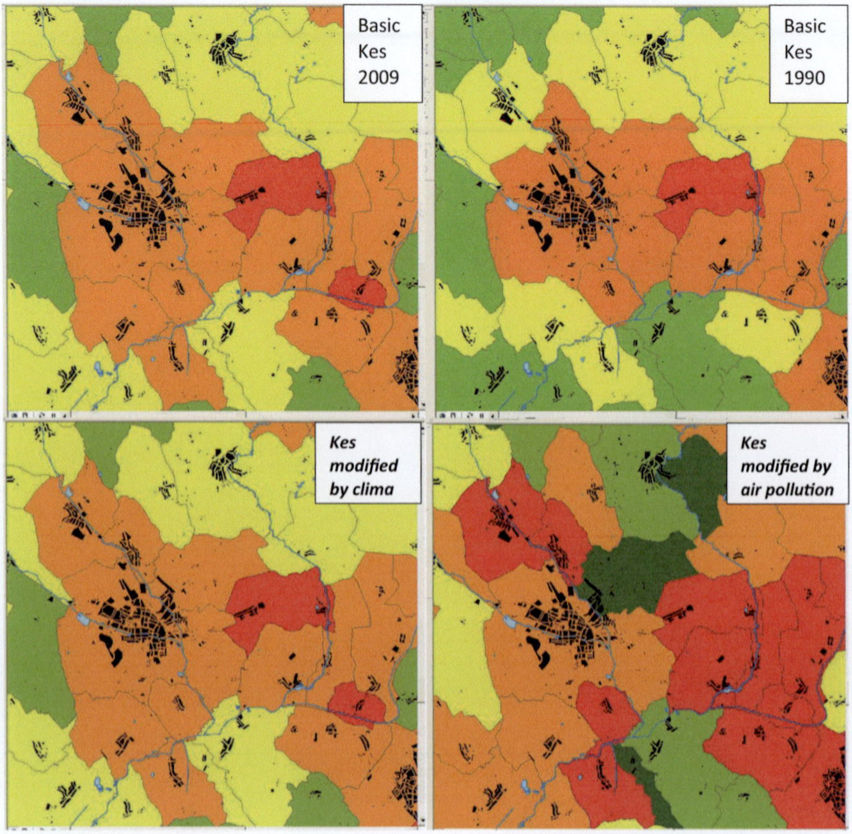

Fig. 5.9 Comparisons of various modified ecological quality values of Lučenec surroundings

$$\boldsymbol{H} = \sum_{i=1}^{n} -\left(\boldsymbol{p}_i \cdot \mathbf{log}_2 \boldsymbol{p}_i\right),$$

where:

$\boldsymbol{p}_i$ probability of the occurrence of individual areas **i** within the entire examined calculated as follows:

$$\boldsymbol{p}_i = \frac{\Delta \boldsymbol{p}_i}{\boldsymbol{P}};$$

$\boldsymbol{\Delta p}_i$ extent of individual areas of LEC;

$\boldsymbol{P}$ the area of the evaluated model territory (e.g. catchment area, geomorphological unit, district, cadastral territory, protected area).

Note: It should be noted that the proportion of each area to the total area is counted separately. Thus, entropy is not merely a sum of areas and their proportion to the total area, but the sum of negative logarithms of proportions of each area to the total area, and therefore a sum of partial entropies of each $\boldsymbol{\Delta p_i}$*!*

The basic landscape-ecological interpretation of individual kinds of entropies is as follows:

- degree of entropy can be regarded as a degree of spatial landscape diversity,
- the more small-scale areas are located in the examined territory, the degree of disarrangement—diversity—disorderliness of the landscape higher.

The basic formula of entropy can provide various procedures, which could meet the needs of landscape-ecological planning. The crucial landscape-ecological factor of correct interpretation is an adequate definition of the territory $\boldsymbol{P}$ and examined elements $\boldsymbol{\Delta p_{i.}}$ The calculations of the individual kinds of entropy are performed utilising the same aforementioned basic formula. The difference can be observed only in the input $\boldsymbol{p_i}$ based on the type of entropy calculation.

For common landscape-ecological studies it is advisable to conduct calculations for the maximum entropy as a comparative base, real entropy, and the entropy according to selected types of CLS elements, as follows:

(a) $\boldsymbol{H_{\max}}$—maximum entropy

$$\boldsymbol{H_{\max}} = \sum_{i=1}^{n} -(\boldsymbol{p_i} \cdot \mathbf{log_2} \boldsymbol{p_i}),$$

for which

$$\boldsymbol{p_i} = \frac{\Delta \boldsymbol{p_{avg}}}{\boldsymbol{P}} = \text{constant value for a given territory; where}$$

$$\Delta \boldsymbol{p_{avg}} = \frac{\boldsymbol{P}}{\boldsymbol{n}} = \text{the average area of LEC areas in a given territory;}$$

$\boldsymbol{n}$ the number of all LEC areas situated within the examined territory.

The value $\boldsymbol{H_{\max}}$ expresses the theoretically maximum possible degree of disarrangement of the territory providing that all LEC areas in the entire territory were, at a given number of areas, of the same size. This value serves as a solid basis for estimation of how much is the current territory less segmented—calculated by basic formula as real entropy $\boldsymbol{H_{real}}$ compared with the maximum possible segmentation that was calculated, i.e. by comparison of $\boldsymbol{H_{\max}/H_{real}}$.

It is illustrated on the example of the model territory of the Ilijský creek catchment (Špinerová 2011). In this case, the area of the whole river basin is 15,605,917 m^2, and the number of LEC areas is 2596, so the average area size is 6012 m^2. Then the formula for the calculation of entropy yields the maximum entropy of the territory $\boldsymbol{H_{\max}} = 11.34$. The maximum entropy is calculated for the entire territory as one single value, and therefore it is a comparative basis for other types of entropy.

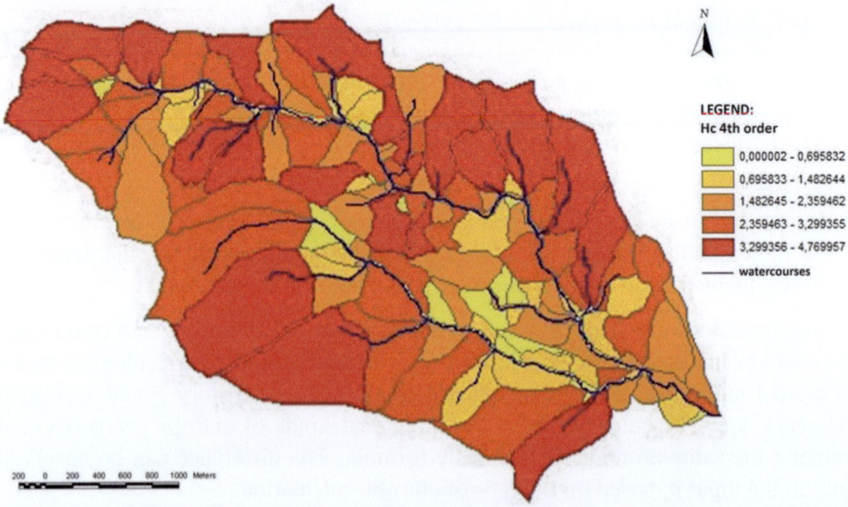

Fig. 5.10 The value of the entropy $\boldsymbol{H}_{\mathbf{real}}$ of microbasins of the model territory. *The darker colour the higher entropy*

(b) $\boldsymbol{H}_{\mathbf{real}}$—the current entropy, where the input data are real LEC areas,

$$p_i = \frac{\textbf{LEC areas} = \Delta p_i}{P}$$

The value $\boldsymbol{H}_{\mathbf{real}}$ represents a real degree of spatial disruption by LEC areas, whereas all areas are considered equal in terms of their content. It is a purely spatial perspective. In general, the greater ruggedness of the examined territory is, the greater landscape-ecological diversity (Fig. 5.10). Distribution of these values in maps has greater explanatory power than just numbers represented in a map, especially their juxtaposition with maximum entropy.

(c) $\boldsymbol{H}_{\mathbf{CLS}}$—entropy by the selected types of CLS (SKŠ) elements, where

$$p_i = \frac{\Delta p_{i\,\mathrm{SKS}}}{P};$$

where

$\Delta p_{i\,\mathrm{SKŠ}}$ the size of individual areas of selected CLS kind in the studied territory.

This value is calculated for partial entropy of the elements of non-forest woody and shrubbery vegetation or the elements linked to forests—the "green" CLS elements. The favourable state is if the evaluated micro-catchment has a high degree of distribution of the "green" elements, which is a prerequisite for establishing areas

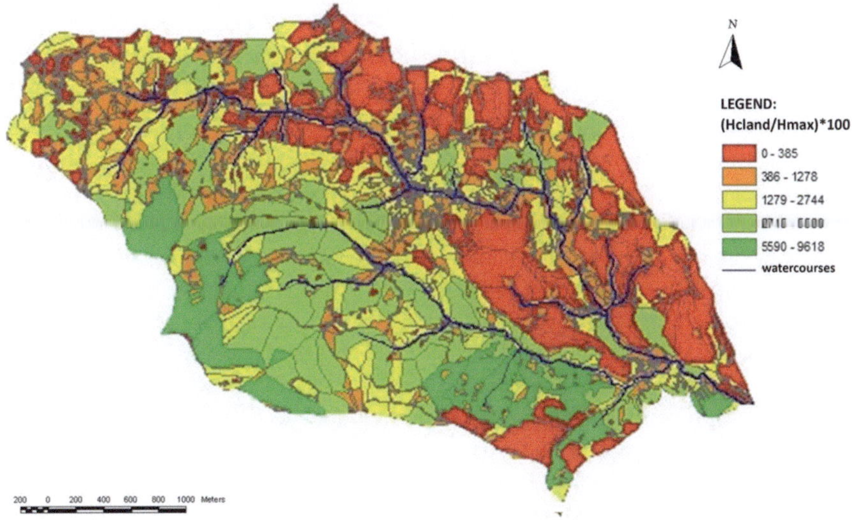

Fig. 5.11 The value of entropy $\boldsymbol{H}_{\mathbf{CLand}}$ caused by the diversity of greenery in comparison with $\boldsymbol{H}_{\mathbf{max}}$ *The higher the value (yellow and green colours) the closer is the real diversity to the maximum ones*

which could further promote biodiversity, development of forest communities, establishing of infiltration zones for surface runoff and hamper the movement of water and transfer of material downhill.

The value of $\boldsymbol{H}_{\mathbf{SKS}}$ expresses the contribution of the green CLS elements to the overall disarrangement of the territory (Fig. 5.11), especially in comparison with the maximum entropy $\boldsymbol{H}_{\mathbf{max}}$.

5.3 Socio-Economic Criteria—Realization Criteria

These criteria are based on the evaluation of the possibilities of optimum land use for specific activities based on a degree of current anthropogenic influence on the landscape, threat to the environment by human activities and legal protection of nature and natural resources. They are derived from the evaluation of socio-economic factors/phenomena (SEF) and their conflicts in the landscape.

5.3.1 *Methodological Aspects of Interpretations of Socio-Economic Phenomena in the Landscape*

In practice, the interpretation of SEF is in many cases expressed in a binary manner—either they occur in a given area or not (see the overview of SEF in the previous chapter). Therefore, their functional values at the site of their manifestation are in the form of restrictions, directives and bans of various kinds with respect to other activities. They are established by legislation or convention, which applies to the SEF of the protection of nature and natural resources as well as various sanitary, protective and safety zones of technical activities, but also other sector zones or administrative boundaries.

As they condition implementation of activities in a given territory they are referred to as **realisation criteria**. Implementation criteria are **derived from the elements of the tertiary landscape structure**.

The values of the SEF of landscape deterioration are a special case. As they have an unstable and transient spatial and temporal character, it is not clearly identified which sustained activities can and cannot be performed in this affected area. The occurrence of the phenomena has rather a limiting impact on the activity itself causing deterioration (e.g. restriction of polluting production). Therefore—paradoxically—although these phenomena have limit values strictly defined (e.g. emission limits, classes of water pollution), for the purposes of landscape engineering i.e. for the proposition of optimal spatial organisation and localisation of human activities in the landscape, they are evaluated as "non-normative", often by means of subjective judgement of a planner, who decides if the landscape in a given territory is "overloaded" or "affected" and curbs implementation of any new technological activities.

In specific cases the activities can cause critical pollution situations, e.g. smog alarms, danger of radiation, critical water pollution; however, they do not represent stable indicators relevant for landscape planning.

Socio-economic phenomena do not need to be specially interpreted. They are sufficiently defined in laws, directives, methodology and methods and their evaluations are summarized in numerous publications (Izakovičová et al. 2000; Miklós et al. 2011). At this point, we will present only the most fundamental aspects of this interpretation.

The interpretation and subsequent classification aims to determine:

(a) **degree of nature conservation** provided by socio-economic phenomena of protection of nature and natural resources and the environment, so called **positive phenomena, priorities for the landscape** (Izakovičová et al. 2000; Hrnčiarová et al. 1997). Comprehensive landscape protection by the action of positive SEF is in LANDEP project spatially represented in form of the territorial system of ecological stability TSES.

(b) **degree of landscape load** by the action of socio-economic phenomena with the nature of stress factors. It is necessary to assess the impact of each SEF separately as well as a degree of the comprehensive load in the landscape by stress factors of primary or secondary point, line and patch type. The output is determination of various degrees and combination of current load of the territory. The load of stress factors is spatially expressed by the territorial system of stress factors, TSSF. The frame of the TSSF comprises of cores, lines—corridors and large areas of TSSF.

These two territorial systems, namely TSES and TSSF, collide, overlap and cause conflicts in the landscape, which in the LANDEP methodology is referred to as

conflicts of interests in the landscape (Izakovičová 1995, 2000).

5.3.2 *Interpretation of the Protection Degree of the Landscape*

The synthesis of positive socio-economic factors/phenomena (SEF) is a spatial projection of the borders demarcating real areas of nature protection, protection of natural resources, cultural monuments and the important elements of CLS. The mapping units on the hierarchical level of TSES, character of the territory and projection scale. The combinations of positive elements act as limiting materials for proposed activities and are crucial for proposals of ecostabilising activities and TSES elements.

Synthetically, mapped SEF can be expressed by superposition, which yields spatial complexes of SEF that can be further interpreted as a degree of nature conservation—**the landscape priority** (Izakovičová et al. 2000). They are specified by implementation prerequisites for required social activities by means of bans, restrictions and supporting factors that are legislatively bound with individual SEF.

An example of the mapping units of positive SEF is in Table 5.11. Admittedly, a certain territory will contain only some of the aforementioned categories, some none, and in some the occurrence of one SEF category may condition the occurrence of the other.

The combinations of positive elements act predominantly as limiting **implementation prerequisites** for production activities, and cause many conflicts of interests in the landscape.

5.3.3 *Interpretation of the Degree of Load on the Landscape*

This interpretation can be attained by spatial synthesis of stress (negative) factors. The spatial synthesis of negative phenomena can be obtained by superimposing individual SEF projections onto a single map. Mapping units are usually divided as follows (Izakovičová et al. 2000; Izakovičová and Moyzeová 2011; Miklós et al. 2011):

- primary stress factors, whose negative impact is their mere physical presence in the landscape, causing barrier effects. The primary stress factors are represented

Table 5.11 Mapping units of positive socio-economic phenomena of local TSES

Group of SEF	SEF—mapping units
Protected areas for nature (Territorial protection)	Protected Landscape Area (PLA)
	National Park (NP)
	Protection zone of NP (PZ NP)
	Zones of PLA and NP
	Protected Area (PA)
	Protection zone of PA (PZ PA)
	Nature Reservation (NR) and National NR
	Nature Monument (NM) and National NM
	Protection zone of NR and NNR
	Protection zone of NM and NNM
	Private protected areas
	Proposed protected nature areas of all level
NATURA 2000	Special Areas of Conservation—SCI—Sites of Community Importance
	Special Protection Areas—PBA—Protected Bird Areas
Protected species	Protected plants
	Protected animals
	Protected minerals
	Protected fossils
Tree protection	Protected trees (PT)
	Buffer zone of PT
Protection of the territory under international conventions	Wetlands Convention (Ramsar sites)
	World Cultural and Nature Heritage sites
	Biospheric Reserves of UNESCO MaB
Elements of TSES in documents	Biocentre of TSES
	Biocorridor of TSES
	Interactive TSES element
	TSES elements of degree
Significant landscape elements (Nature and Landscape Conservation Act)	Gene pool sites of flora
	Gene pool sites of fauna
	Significant landscape structures
Protection of forest resources	Protective forests, as: • Forests in exceptionally unfavourable places • High mountain forests, • Forests in the dwarf pine belt Other forests with predominant protection of soil

(continued)

Table 5.11 (continued)

Group of SEF	SEF—mapping units
	Special-purpose forests, as: • Forests in protective zones of water resources, • Forests in nature conservation areas, • Forests designated for research • Forests with medical-recreational function, • Suburban forests with health-recreational function • Forests in recognized animal and pheasant grounds • Forests under the influence of emissions
	Timber production forests
	Military forests
Protection of water resources	Protected water area (protected area of natural accumulation of water)
	Water-supply rivers and reservoirs
	Catchments of water-supply rivers and reservoirs
	Important water flows
	1st to 3rd zones of hygiene protection of underground and surface water
Protection of natural healing resources	Natural healing water sources
	Natural healing sources of gases and emanations
	Natural healing resources from peat fens, sludges and other earth healing sources
	1st–3rd protective zones of sources of mineral water
	Natural sources of spring waters
	1st–3rd protective zones of sources of spring waters
Protection of natural healing spas and climatic spas	Spa cities (natural healing spa)
	Interior spa territory
	Exterior spa territory
	Protective zone of the spa territory
	Climatic zones beneficial for healing
Protection of soil resources	High quality soil (with the best production capability in the relevant cadastral area)
	Specially protected soils (with hydromeliorating and other measures, such as orchard, vineyard etc.)
Protection of gene pool	Protected fishing areas
	Fish-farming flows
	Recognized animal keepers
	Independent pheasantries
Protection of mineral resources	Protected area of state owned mineral deposits
	Protected area of non state-owned mineral deposits

(continued)

Table 5.11 (continued)

Group of SEF	SEF—mapping units
Protection of the monuments	Cultural monuments
	National cultural monuments
	Urban conservation monuments
	Folk Architecture Monument
	Urban monuments zones
	Rural monuments zones
	Archaeological find
	Protection zones of the memorial fund

by material objects of the elements of the current landscape structure, thus their spatial projection is identical with the objects that are obtained in the processing of the CLS;
- secondary stress factors, including either intangible phenomena, such as various declared protection, hygienic and safety zones;
- deterioration phenomena—pollutions of the elements of primary and secondary landscape structure although unstable their spatial incidence is also mapped.

A key of the mapping units of stress factors of a local TSES is exemplified in Tables 5.12 and 5.13.

A comprehensive spatial projection of all stress factors occurring in the concerned territory and the extent of their negative impact are represented in **the territorial system of stress factors TSSF**, which also denotes a continuous barriers for the elements of territorial system of ecological stability.

The output of the syntheses is delimitation and re-evaluation of areas with different combination of stress factors.

The territories containing the combinations of stress factors are subsequently individually assessed according to the types and intensity of the impact of the stress factors and degree of load on the territory. Based on the combination of stress factors, certain typological territories of stress factor combinations can be identified (Izakovičová et al. 2000):

- territories with multiple and diverse stress factors
- territories with the accumulation of several diverse stress factors
- territories with one-two dominant stress factors resulting from industrial development.

Spatial interpretation of the spatial system of stress factors TSSF is as follows:

(a) **The cores of TSSF** are point or small-scale objects, where stress factors accumulate, and linear elements meet or intersect. The cores may be surrounded by large-scale areas of SEF with negative impact, which are directly linked with the cores such as intensively cultivated and fertilised arable land around the

Table 5.12 Mapping units of primary stress factors on local level

	Objects and areas
Industrial and mining objects and areas	Plant sites (industrial sites, warehouses, etc.)
	Objects of surface mining
	Objects of subsurface and underground mining
	Industrial waste landfills Sludge beds
	Undermined areas
Energy pipelines and other pipelines	Very high voltage power lines
	High voltage power lines
	Low-voltage power lines
	High-pressure gas pipelines
	Medium pressure gas pipelines
	Low pressure gas pipelines
	Steam pipeline
	Oil pipelines
	Sewerage system
	Water supply system
	Cable lines
	Substations
Transportation	Highways
	Very heavily loaded road corridors
	Heavily loaded road corridors
	Medium loaded road corridors
	Very heavily loaded railway corridors
	Heavily loaded railway corridors
	Medium loaded railway corridors
	Cog railway
	Railway areas
	Parking areas
	Airport areas
Agricultural objects and areas	Areas of agricultural objects
	Field airports
	Field storages of fertilizers
	Intensive agricultural production (arable lands with big blocks)

(continued)

Table 5.12 (continued)

	Objects and areas
	Garden villages with objects
Objects and areas of forestry and water management	Forestry technologies (warehouses, transport technologies, mechanization yards)
	Water management structures
	Water management structures and objects
	Regulated streams and water reservoirs
Residential areas	Residential areas with the prevalence of public services
	Residential areas with block houses
	Residential areas with family houses
	Individual buildings outside of the residential areas
	Construction sites
Objects and areas of sports and recreation	Cottages and garden settlements, fishing huts
	Campsites, public camps
	Spa and healing areas
	Swimming pool areas
	Sports fields
	Riding area
	Ski lift chair and lift cabin
	Ski lifts
	Ski springboard
	Tobbogan run
	Ski trails downhill
	Cross-country skiing
	Hiking and educational walkways
	Cycling routes
	Waterways
	Ports for water sports
Other elements	Border Guard Objects
	Municipal waste disposal
	Non agricultural land, abandoned areas
	Devastated areas

Table 5.13 Mapping units of secondary stress factors on local level

Source of pollution	Characteristics
Sources of air pollution	Sources of air pollution—big, medium, small,
	Sources of air pollution—mobile
Sources of water pollution	Sources of water pollution—big, medium, small
Sources of noise	Sources of noise—big, medium, small
Source of radiation	Source of radiation
	Phenomena of deteriorization—secondary stress factors
Polluted air	Air pollution—big, medium, small
Environmental noise	Environmental noise—big, medium, small
Polluted waters	Polluted watercourses, heavily, very heavily
	Polluted groundwater, heavily, very heavily
Damaged soils	Soil risks
	Contaminated soil, heavily contaminated soil
	Soil erosion damage—very heavy, heavy, medium
Damaged vegetation	Damage to vegetation—medium, strong
	Damage to vegetation—strong
	Zones of hygiene protection (ZHP) and safety zones
Areas ZHP	Industrial sites ZHP
	Landfill sites ZHP
	Wastewater treatment plants ZHP
	Agricultural objects ZHP
Safety zones	Safety zones of railways
	Safety zones of roads
	Safety zones of power lines
	Safety zones of gas lines
	Safety zones of cable lines
Protected zones	Military zones
	Prírodné stresové faktory
Threatened territories	Areas threatened by landslides*
	Areas threatened by avalanches*
	Areas with increased seismicity*
	Areas with increased natural radiation*
	Flood plains

*areas can be classified based on the degree of risk exposure of the given stress factor

agricultural technical areas, areas of air pollution, soil contamination around industrial buildings, water contamination around landfills etc.

At the local level, the TSSF cores are typically represented by a point or small-scale objects, such as industrial areas, sewage treatment plants, storage areas and other technical works, landfills and settling pits, mining areas, traffic areas, agro-technical buildings and areas, livestock production farms, residential areas of private house or housing estate character, administrative and commercial areas, recreational and sports areas, cottage settlements, garden allotments etc.

At the regional level, the TSSF cores are considered the areas of residential units, larger areas of industrial and technical objects which can be clearly spatially identified and mapped.

Essentially, the special-purpose core classification is based on the evaluation of the intensity of their stress impact. As for the primary influence, the main classification criterion is the size of the area and the intensity of the barrier effect. The secondary influence applies the classification criterion expressed by the type and intensity of secondary influence—emission production, noise, dust, bacteriological source of pollution etc. The combination of these criteria allows the classification of the TSSF cores into relative scales from the most to the least loaded.

(b) **Linear elements—the TSSF corridors** represent anthropogenic lines—roads, railways, power lines, pipelines or adversely altered and affected natural linear elements—pollute watercourses, canals. The evaluation of these stress factors should approach each stress factor individually and in case of more parallel lines, their barrier and deteriorative cumulative effect must be taken into consideration. The approach adopted for the assessment of individual lines is fundamentally the same as for the classification of the TSSF cores: the surface extent and intensity of secondary influence is assessed. Watercourses are evaluated based on the pollution level of surface water, transport corridors on the intensity of transport, power lines and other pipelines on the installed capacity, or the width of their protection zones. The result indicates a degree of stress impact of lines on the landscape or a degree of the load of TSSF lines arranged from the least to the most loaded ones.

(c) **Large-scale areas of TSSF** are mostly linked with the cores and typically form zones with a cumulative effect of various stress factors adjacent to individual cores. This is related particularly to the accumulation of load of the territory because of secondary stress factors. The assessment should be conducted based on the intensity of stress factor effect present in the evaluated territory (Fig. 5.12).

The interpretations of the protection degree of the landscape and the load of the territory employ a wide range of various methods. Often a formalized, semi-quantitative methods is applied, for example, to calculate the sum of ecological significance of the landscape based on positive SEF, or the intensity of secondary stressors in a particular territory or to calculate their weighted sum, while the weighting coefficients are determined according to the relative importance of a stress factor. The results can be categorized into relative scales and thus classify the areas based on

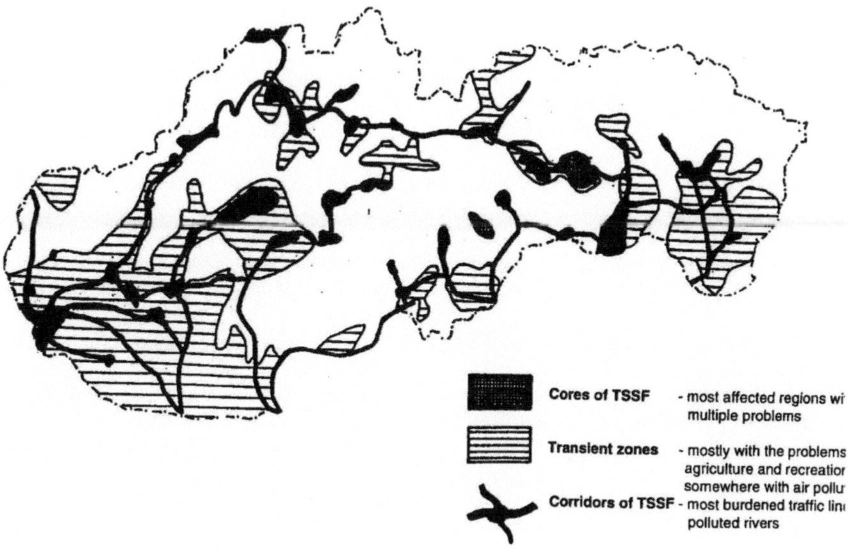

Fig. 5.12 Scheme of the territorial system of stress factors TSSF in Slovakia

their protection level or load level from the non-loaded areas—without stress factors or only local occurrence of stress factors—to extremely loaded—territories with the values of stress factors exceeding the acceptable limits (e.g. Hrnčiarová et al. 1997; Moyzeová and Izakovičová 2009; Izakovičová et al. 2011).

References

Chlapečka M (2004) Výpočet potenciálnej erózie pôdy pomocou geografických informačných systémov. Bioklimatické pracovné dni 2004, 23–26 Aug 2004, Viničky. Elektronický zdroj

De Roo A, Thielen J, Gouweleuuw B (2002) LISFLOOD, a distributed water-balance, flood simulation and flood inundation model. User Manual Version 1.0, Ispra (I): Joint Research Centre JRC

Diviaková A (2010) Hodnotenie líniových formácií nelesnej drevinovej vegetácie pre potreby územných systémov ekologickej stability. VKÚ, a.s, Harmanec

Diviaková A (2011) Biotické komplexy pre environmentálny manažment. VKÚ, a. s, Harmanec

Fu P, Rich PM (2002) A geometric solar radiation model with applications in agriculture and forestry. Comput Electron Agric 37:25–35

Halada Ľ (1998) Krajinnoekologické hodnotenie vegetácie. ÚKE SAV Nitra, Kandidátska dizertačná práca

Halada Ľ, Topercer J, Mederly P, Kartusek V (1995) Systém ekologickej kvality krajiny – ďalší prístup k manažmentu krajiny. Životné prostredie 29(5):271–273

Hreško J (1996) Abiotické predpoklady výskytu ekologicky hodnotných biotopov pre územie Slovenska. In: Hrnčiarová T (ed) Celostnosť - syntéza - ochrana životného prostredia. (zborník referátov pri príležitosti 30. výročia založenia ÚKE SAV), Bratislava: Ústav krajinnej ekológie SAV

Hrnčiarová T (1999a) Prepojenie metodiky LANDEP na metodiku EÚK. Životné prostredie 33(1):11–16

Hrnčiarová T (1999b) Krajinnoekologické plánovanie pomocou metodiky LANDEP a metodiky EÚK. Geografický časopis 51(4):399–413

Hrnčiarová T (2001) Ekologická optimalizácia poľnohospodárskej krajiny (modelové územie Dolná Malanta). Veda SAV, Bratislava

Hrnčiarová T (2009) Indikátory hodnotenia zmien diverzity krajiny. In: Pucherová Z, Vanková V (ed) (2009) Problémy ochrany a využívania krajiny - teórie, metódy a aplikácie. Nitra, Združenie Biosféra. ISBN 978-80-968030-9-5. pp 133–142

Hrnčiarová T, Miklós L et al (1997) Ekologická únosnosť : metodika a aplikácia na 3 benefičných územiach. I.-IV. Bratislava : Ústav krajinnej ekológie SAV a Ministerstvo životného prostredia SR

Hrnčiarová T et al (1996) Ekologická únosnosť krajiny: metodický postup. Etapová správa. ÚKE SAV, Bratislava

Izakovičová Z (1995) Ecological interpretations and evaluation of encounters of interests in landscape. Ekológia (Bratislava) 14(3):261–275

Izakovičová Z (2000) Evaluation of the stress factors in the landscape. Ekológia (Bratislava), 19(1):92–103

Izakovičová Z et al (2000) Metodické pokyny na vypracovanie projektov regionálnych ÚSES a miestnych ÚSES. Združenie krajiny 21, Bratislava

Izakovičová Z, Moyzeová M (2011) Hodnotenie významnosti a zaťaženia územia pre potreby územných systémov ekologickej stability. In: Ekologické štúdie: Vedecký časopis venovaný aktuálnym problémom ekológie, krajinnej ekológie a príbuzných vedných disciplín, roč. 2, č. 1–2:18–26

Izakovičová Z, Moyzeová M, Oszlányi J (2010) Problems in agricultural landscape management arising from conflicts of interest—a study in the Trnava Region, Slovak Republic. Innovations in European rural landscapes. Springer, Heidelberg, pp 77–96

Izakovičová Z, Miklós L, Moyzeová M et al (2011) Model geoekosystémov na regionálnej úrovni. Ústav krajinnej ekológie SAV, Bratislava

Jurko A (1986) Krajinnoekologická významnosť rastlinných spoločenstiev. In: Ekologická optimalizácia využívania Východoslovenskej nížiny. III. diel. ÚEBE SAV - Slovosivo Michalovce, pp 214–218

Jurko A (1990) Ekologické a socioekonomické hodnotenie vegetácie. Príroda, Bratislava

Kalivodová E, Kminiak M (2003) Krajinnoekologické metódy hodnotenia biotických zložiek krajiny s dôrazom na regionálnu úroveň. In: Kozová M, Bedrna Z (eds) Krajinnoekologické metódy v regionálnom environmentálnom hodnotení, pp 114–118

Kerényi A, Csorba P (1991) Assessment of the sensitivity of the landscape in a sample area in Hungary for climatic variability. Earth Surf Process Land 16:663–673

Kim HS (2006) Soil erosion modeling using RUSLE and GIS. Colorado State University, Fort Collins (US)

Kočický D et al (2008) Komplexný program protieróznej ochrany a návrh opatrení na zvýšenie retenčnej schopnosti územia SR v členení podľa čiastkových povodí. Záverečná správa. Esprit s.r.o, Banská Štiavnica

Kontriš J (1978) Fyziognomicko-ekologická typizácia lesov a krovín a ich funkcia v krajne. In:Problémy ekológie krajiny, Bratislava, 23:81–122

Kozová M, Bedrna Z (ed) (2003) Krajinnoekologické metódy v regionálnom environmentálnom hodnotení. Univerzita Komenského v Bratislave

Löw J et al (1995) Rukověť projektanta místního ÚSES. Agroprojekt, Brno

Maglocký Š (1993) Hodnotenie biotopov. Životné prostredie 27(4):196–199

Mander Ü, Müller F, Wrbka T (2005) Functional and structural landscape indicators: upscaling and downscaling problems. Ecol Ind 5(4):267–272

Mariot P (1983) Geografia cestovného ruchu. Veda, Bratislava

McGarigal K (2002) Landscape pattern metrics. In: El-Shaarawi AH, Piegorsch WW (eds) Encyclopedia of Environmetrics, vol 2, Wiley, Chichester, pp 1135–1142

Míchal I (1992) Ekologická stabilita. Veronica, Brno

Midriak R (1997) Ohrozenie povrchu krajiny – limit jej trvaloudržateľného rozvoja. In: Midriak R (ed) Trvaloudržateľný rozvoj krajiny. Zbor. ref., TU Zvolen, pp 21–26
Midriak R (1999) Ekologická únosnosť vysokohorských oblastí z hľadiska morfodynamických procesov. Bratislava, ÚKE SAV. Životné prostredie 33(1):42–45
Miklós L et al (1986) Ekologický plán VSN. Súbor grafických výstupov. Ekologická optimalizácia využívania VSN. IV. diel. ÚEBE SAV Bratislava - Slovosivo
Miklós L (1991) Zásady slovenskej ekologickej politiky. Bratislava, Životné prostredie 25(4):174–178
Miklós L (1992) Protecting the Environment during Regional Development. In: Vasko T (ed) Problems of economic transition. Regional development in Central and Eastern Europe. Chap 5. AVEBURY, Aldershot—Brookfield USA—Hongkong—Singapore—Sydney, pp 51–62
Miklós L, Krcho J, Hrnčiarová T, Matečný I, Kozová M (1997) Morfometrické ukazovatele reliéfu a ich účelové interpretácie pre plánovacie procesy. Nadácia Katedry UNESCO pre ekologické vedomie, Banská Štiavnica
Miklós L, Diviaková A, Izakovičová Z (2011) Ekologické siete a územný systém ekologickej stability. Vydavateľstvo TÚ vo Zvolene
Miklós L et al (2011) Geografický informačný systém povodia Ipľa: katalóg GIS a výber máp. Bratislava: Ústav krajinnej ekológie SAV, Katedra UNESCO. Fakulta ekológie a environmentalistiky, Technická univerzita vo Zvolene: Esprit Banská Štiavnica
Moyzeová M, Grotkovská L (2006) Environmental aspects of wetland evaluation (model area of the Paríž water stream). Ekológia (Bratislava) 25(1):169–178
Moyzeová M, Izakovičová Z (2009) Strety záujmov medzi ochranou a využívaním krajiny. In: Pucherová Z, Vanková V (ed) Problémy ochrany a využívania krajiny – teórie, metódy a aplikácie.Zborník vedeckých prác. Združenie Biosféra, Nitra, pp 263–268
Németh R, Verrasztó Z, Izsák T (2016) A mérnökgeológia feladatai és lehetőségei a környezet védelmében – környezeti és társadalmi konfliktusok Aknaszlatinán. Mérnökgeológia, Kőzetmechanika, pp 67–82
Odum EP (1977) Základy ekologie (Fundamentals of Ecology—Czech translation). Academia, Praha
Papánek F (1974) Čo sú funkcie lesa? Les 2:61–70
Papánek F (1978) Teória a prax funkčne integrovaného lesného hospodárstva. Lesnícke štúdie 29, VÚLH Zvolen, Príroda, Bratislava
Pauditšová E (2010) Štúdia vizuálneho vplyvu veterných parkov na krajinnú scenériu. In: Krajinné plánovanie. Bratislava, STU, pp 238–257
Rich PM, Dubayah R, Hetrick WA, Saving SC (1994) Using viewshed models to calculate intercepted solar radiation: applications in ecology. American Society for Photogrammetry and Remote Sensing. Technical papers, pp 524–529
Ružička M, Hrnciarová T, Miklós L (ed) (1988) Spatial and functional relationships in landscape ecology. VIIIth International Symposium on Problems of Landscape ecological Research, vol 1. IEBE CBES of SAS. IALE East European Region. Bratislava
Špinerová A (2011) Prvky využitia zeme ako regulátor zón dynamiky svahu v modelovom území Ilijského potoka. Doktorská dizertačná práca. TU Zvolen
Špinerová A (2015) Štruktúra krajiny ako regulátor dynamiky pohybu vody a materiálu. Zvolen, Vydavateľstvo TU vo Zvolene
Špulerová J (2010) Evaluation of vegetation and their limits for sustainable development. GeoScape: Altern Approaches Middle-Eur Geogr 5(1):175–183
Špulerová J, Dobrovodská M, Izakovičová Z, Kenderessy P, Petrovič F, Štefunková D (2013) Developing a strategy for the protection of traditional agricultural landscapes based on a complex landscape-ecological evaluation (the case of a mountain landscape in Slovakia). Moravian Geogr Rep 21(4):15–26
Šteffek J et al (1992) Biologicko-ekologické kritériá územného systému ekologickej stability. Ekotrust, Banská Štiavnica
Štefunková D, Cebecauer T (2006) Visibility analysis as a part of landscape visual quality assessment. In: Ekológia (Bratislava): Int J Biosph 25(1):229–239

Szabó SZ, Csorba P, Varga K (2008) Landscape indices and landuse—tools for landscape managment. In: Plit J, Andreychouk V (eds) Methods of landscape research. Dissertations Comm Cult Landscape 8:7–20, ISSN 1896-1460, Sosnowiec, Poland

Szabó SZ, Csorba P, Szilassi P (2012) Tools for landscape ecological planning—scale and aggregation sensitivity of the contagion type landscape metrics indices. Carpathian J Earth Environ Sci 7(3):127–136

Szeicz G, Long IF (1969) Surface resistance of crop canopies. Water Resour Res 5:622–633

Tremboš P (1994) Krajinnoekologické podmienky a regionálny rozvoj. In: Trvaloudržateľný rozvoj a krajinnoekologické plánovanie v európskych horských regiónoch. TU vo Zvolene, pp 163–166

Turner MG, Gardner RH (eds) (1991) Quantitative methods in landscape ecology: the analysis and interpretation of landscape heterogeneity. Springer, Berlin

Urbánek J (2009) Geomorfologická katéna. Veda, Bratislava

Verrasztó Z (1979) Land formation and the geological aspects of environmental protection. In: Symposium changes of the geological environment under the influence of man´s activity. IAEG National group, Krakow-Sandomierz Belchatow-Plock-Warszawa, pp 135–141

Wischmeier WH, Smith DD (1978) Predicting rainfall erosion losses—a guide to conservation planning agriculture handbook No. 537. Washington D.C., US Department of Agriculture

Zachar D et al (1989) Polyfunkčná zeleň v poľnohospodárskej krajine. Slovenská bioklimatologická spoločnosť, Bratislava, pp 98–130

Zaušková Ľ (2001) Ekologická únosnosť lesnej krajiny a ochrana pôdy v gravitačnom území VN Hriňová v biosférickej rezervácii Poľana. Dizertačná práca, FEE TU, Banská Štiavnica

Chapter 6
Landscape-ecological Evaluation in LANDEP

Abstract The evaluation is the process of determining the suitability of geocomplexes for the localisation of selected activities and instituting optimisation measures. Evaluation is a **confrontation** of requirements of individual activities for certain landscape properties with actual values of these properties assisted by the criteria, and arguments produced by formal or informal interpretation of the landscape properties. Basic step of the evaluation process is the definition of functional values of the indicators—the **degree of suitability** of indicators for selected activities. In the LANDEP project, we recognise:

- **excluding** values—for practical purposes sometimes marked as **zero values,** i.e. the values which, considering current conditions, both from technological and ecological point of view, make performance of a given activity impossible;
- **Limiting**—threshold, restrictive values which are considered unsuitable from the landscape-ecological perspective although from the technological point of view, they are not necessarily perceived as limits;
  If the values range between limit and zero values for current use, such a use needs to be **changed** and a not limited activity should be applied;
- **Conditionally unsuitable/suitable**—these values allow a certain activity, but should not be 'sacrificed' for it considering the overall strategy and landscape-ecological perspectives;
- **Suitable**—other (over-limit, unlimited) values. These include values which are considered to make the performance of a given activity possible. They can be categorised into degrees of suitability ranging from the most suitable to the least suitable.

If the values range between limit and zero values for current use, such a use needs to be **changed** and a not limited activity should be applied.

- **Conditionally unsuitable/suitable**—these values allow a certain activity, but should not be 'sacrificed' for it considering the overall strategy and landscape-ecological perspectives;
- **Suitable**—other (over-limit, unlimited) values. These include values which are considered to make the performance of a given activity possible. They can be

L. Miklós and A. Špinerová, *Landscape-ecological Planning LANDEP*,
https://doi.org/10.1007/978-3-319-94021-2_6

categorised into degrees of suitability ranging from the most suitable to the least suitable.

If a value of current land use lies within the interval of unlimited values, the current form of land use **can be maintained**. It is also acceptable to change it to other non-limited activity. The functional values are arranged into a logic table. In some cases, the evaluation process is amended also by definition of weight coefficients of the functional values for selected activities. For the evaluation in LANDEP a formalised procedure has been developed-the decision process in table. The **decision table** has different logically arranged functional zones allowing synoptical tabular sequence of the confrontation of values of indicators with selected activities. The decision table displays the system of abiocomplexes, landscape-ecological complexes, application of functional values, up to decisions on changes of current land use. The chapter is illustrated by numerous tables and maps.

Keywords Suitability · Functional values · Limits · Decision table

The optimisation process of LANDEP simulates a decision-making process that takes place in the mind with a purpose to achieve the most favourable utilisation of space. There are numerous methods for the evaluation processes depending on the goals of individual studies. Generally, the evaluation of the landscape properties belongs traditionally to one of the main stream of the applied landscape ecology (Barsch et al. 1993; Bastian and Schreiber 1994; Bastian et al. 2006) and this statement is valid up today (Lóczy 2002; Mizgajski and Stepmiewska 2012; Jacobs et al. 2016). Recently, the most massive wave of landscape evaluations concerns the assessment of ecosystem services, also in very diverse ways and forms (Jones et al. 2013; Mederly et al. 2015). This process may be very simply, straightforward, in case it regards a small area or a low number of required activities, but it can become very intricate in other cases. That is why in the LANDEP methodology, the decision-making process was divided into individual phases. In order to maximise systematisation and objectification of the results, some formalised procedures were elaborated within the phases of a decision-making process.

The decision-making process of the LANDEP methodology was basically divided into degrees of evaluations and propositions while we emphasise that the degrees overlap.

The LANDEP methodology defines an **evaluation** as a process of determining the suitability of LEC types for localisation of selected social ***R*** activities and instituting optimisation measures.

Evaluation forms the core of decision-making process: it is a confrontation requirement of individual activities for certain landscape properties with actual values of these properties assisted by the criteria, arguments produced by formal or informal interpretation of the landscape properties.

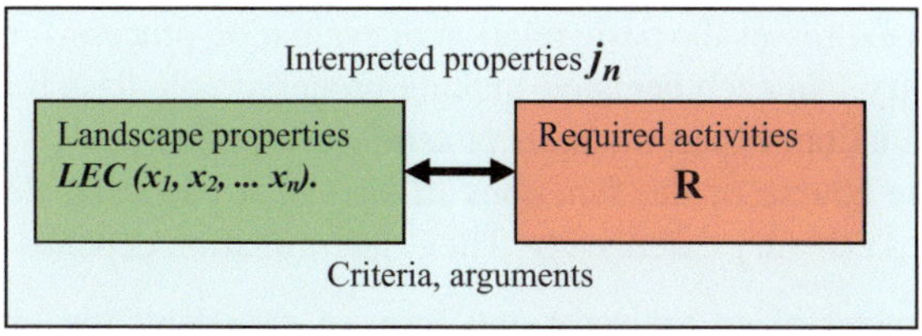

Formalised procedure and logic of decision-making process is as follows:

6.1 Functional Values of the Indicators

The indicators of landscape properties $\boldsymbol{x_n}$, expressed in the form of state variables or typological characteristics, were synthetised into synthetic units of LEC types

$$\mathbf{LEC}[\mathbf{ABC}\{x_1, x_2, \ldots, x_{12}\}, \mathbf{CLS}\{y_1\}].$$

The state variables or typological characteristics (spatial subsystems) are characterised by values which, in their original uninterpreted form, will be regarded as the **absolute values of the indicators** $\boldsymbol{x_n}$ (e.g. relief inclination, soil depth, etc.) as they were characterised in the previous chapters.

If these absolute values $\boldsymbol{x_n}$ are evaluated against individual required social activities $\boldsymbol{R}$, every single individual absolute indicator value $\boldsymbol{x_n}$ will acquire the character of relative value which, to some extent, ensures, restricts or prevents performance of the activity, therefore this new characteristics assigned to the absolute values will be referred to as *functional values* $\boldsymbol{s_{xi}}$ for each required activity $\boldsymbol{R}$. Thus, each absolute value $\boldsymbol{x_n}$ acquires relative importance—**suitability degree, functional suitability**—of a given indicator value for selected activity. This also means that the same absolute value $\boldsymbol{x_n}$ can gain a high functional value—suitability degree—$\boldsymbol{s_{xn}}$ for one activity and at the same time low suitability for the other c.f. high slope inclination (as an absolute value), which is unsuitable for use as arable land, but very suitable for skiing purposes (functional values–functional suitability).

A formal education of the process is as follows: a **suitability degree/functional suitability** is determined for each indicator value $\boldsymbol{x_n}$ for each required activity $\boldsymbol{R}$ as functional value $\boldsymbol{s^R xn}$. In other words, every indicator of properties (state variable or typological characteristic) $\boldsymbol{x_n}$ in every real LEC type has its specific absolute value $x_n^{\mathbf{LEC}i}$. Confronted with the requirements of individual activities $\boldsymbol{R}$, the absolute values $x_n^{\mathbf{LEC}i}$ acquire relative degrees of suitability—functional suitability $\boldsymbol{s^R xn}$ in the geosystem $\mathbf{LEC}_i$ for an activity $\boldsymbol{R}$, i.e.

$$\boldsymbol{s}_{xn}^{\mathbf{LEC}i\,R} = f(\boldsymbol{x}_n^{\mathbf{LEC}i}).$$

This formula expresses the basic relation of evaluation process— the **determination of suitability**—in each decision-making process, regardless if its executed by means of a calculation, expert methods or semi-quantitatively.

Naturally, the course of the functions in case of a particular determination of functional values can vary extensively. Their determination depends on:

- physical character of an indicator and form of representation of their absolute values (c.f. the overview of state variables and spatial subsystems in Chap. 3);
- the form of expressing the requirement of the individual activities for a given indicator (strict requirements numerically expressed or verbally formulated ‘desires’);
- existing methods and confrontation ability, e.g. existing legally established standards (e.g. the system of legally prescribed technical norms), various sectoral and internal norms (e.g. maximal and minimal values), conventions, case studies and expert determination of suitability.

6.1.1 The Character of Functions and Cardinal Functional Values

Functional values can be defined by the criteria—purpose oriented properties j_n—as described in the Chap. 5. The course of functions, which determine functional values, is very diverse. The functions can have continuous course, when each value x_n can be assigned a certain value s_{xn}, be mathematically definable, but often indefinable, or a diverse, almost random course, up to the individual determination of the functional values to each single absolute indicator value.

Specific values s_{xn} for each selected activity R can be ranked from the best to the worst one.

Expression of Functional Values

The tendency of functional values can be entirely different from the tendency of the absolute values, which are mostly in increasing, decreasing, periodical or other arrangement. For instance, the soil granularity values are ordered from the lightest to the heaviest, but in terms of arable land, the granularity values do not exhibit such a tendency—moderately heavy soils are the most suitable for arable soils, and the functional suitability of granularity decreases to both directions until it reaches the limit values (i.e. entirely unsuitable sands and clays). The situation is even more complicated when determining the functional values of biotic indicators (e.g. suitability of vegetation communities for various activities) where degrees of suitability need to be assigned to individual values.

The form of expressing functional indicator values of landscape properties can be very diverse, for example, as:

- financial evaluation (e.g. arability of soil determined by the costs of ploughing);

- physical variables (e.g. engineering, geological bearing capacity of the substrate for building activities, suitability of an area for vineyards based on the amount and length of sunlight), these can be arranged continually or in scales (intervals);
- proportional suitability—'distance' from the worst and the best values (e.g. the calculation of suitability in the LANDEP methodology—Miklós et al. 1986a; Múdry et al. 1995);
- score values (widely used in various landscape plans, e.g. for recreational activities);
- qualitative suitability scales (e.g. the most precious to the least precious ecosystems for nature conservation), or their cardinal forms such as 'high–moderate–low' or 'yes–no', which often leads to unwieldy formulations in order to achieve an accurate verbal representation of intermediate categories.

Generally speaking, in scientific studies, there are also efforts to approximate functional values of landscape-ecological indicators with the form of technical norms and present them in measurable quantities because decisions driven by such evaluation are legislatively recognised and accepted by technicians, planners, managers and executive officers. Other forms for such decisions are often disrespected. Therefore, many authors have attempted to formalise determining of functional values of qualitative indicators. There is a considerable number of landscape-ecological indicators, with varying degrees of success (e.g. Bastian and Schreiber 1994; Múdry et al. 1995; Löw et al. 1995; Hrnčiarová 1999b; Sklenička 2003; Kozová and Bedrna 2003; Kozová et al. 2010). Despite the time lag, the most comprehensive semi-quantitative methods, objectively aimed at further steps of decision-making process, were formalised and formulated in collective works for ecological optimisation of the Eastern Slovak Lowland (Miklós et al. 1986; Kozová et al. 1986) and development of the methodology of ecological carrying capacity and its application in three model territories (Hrnčiarová and Miklós 1997). This approach for the formalisation of the evaluation process is used up today (Hreško et al. 2003, 2008; Špinerová 2010, 2015; Boltižiar 2011; Diviaková 2011; Csorba and Szabó 2012; Špulerová et al. 2013; Mederly et al. 2015).

Cardinal Values

The landscape-ecological plans utilise for definition of functional values s_{xn} most frequently the so-called **cardinal** values and interval values of analytical properties x_n. A categorical binary evaluation type 0–1 (suitable–unsuitable) is also abundantly used.

Specific and frequently used values are those that limit or restrict performance of a given activity. These values are commonly called **limits, limit values**. This value divides the set of values x_n into over-(suitable) and below-limit values.

In the LANDEP project, we recognise:

- **excluding** values—for practical purposes sometimes named and marked as **zero values** (in following tables marked as ***0***)—i.e. the values which, considering current conditions, both from technological and ecological point of view, make performance of a given activity impossible;

- **Limiting**—threshold, restrictive, and **unsuitable** values which are considered unsuitable from the landscape-ecological perspective although from the technological point of view, they are not necessarily perceived as limits (in following tables marked as ***L***). This value has key significance for ecologically optimal utilisation of the territory Areas of LEC acquire values over or below this limit, which means threshold between suitable and unsuitable conditions for a given activity.

Cardinal importance for ecologically optimal land use: if the x_n values range between **limit and zero** values for current use of an area, such a use needs to be **changed** and an activity that is not limited by the values should be applied.

- **Conditionally unsuitable/suitable**—these values allow a required activity, but should not be 'sacrificed' for the required activity from the overall strategy and landscape-ecological perspective—deep soils on flatlands are considered unsuitable for building and construction activities and also forests, not because it is impossible from technological of landscape-ecological point of view, but such areas should be primarily used for intensive agricultural production [in following tables marked as (***L***)];
- **Suitable**—other (over-limited, unlimited) values. These include values which are considered to make the performance of a given activity possible. They can be categorised into degrees of suitability ranging from the most suitable to the least suitable, and subsequently ordered based on their suitability level (in following tables marked by numbers).

Cardinal importance for ecologically optimal land use: if an x_n value of current land use lies within the interval of **unlimited** values, the current form of land use **can be maintained**. It is also acceptable to change it to other, different non-limit activities. However, it should be taken into account that the lower the suitability degree (e.g. closer to the limit values), the more subsequent measures will have to be instituted to maintain the suitability for the current activity.

Selected Aspects of Determining Functional Indicator Values

Determining a degree of functional suitability of limit and zero values of individual indicators x_n for selected activities is based on landscape-ecological principles which should reflect:

- feasibility of technical performance of an activity, i.e. real functional suitability of a given value of an interpreted property (if possible, taking into account applicable regulations, norms or conventions to assess suitability of the activity)—this point resembles the evaluation of landscape properties in projection practice;
- the perspective of protection and development of the ecologically stable landscape, protection of soil resources (anti-erosion function of activities), water sources, biotic functions, etc;
- aspect of reaction of the landscape to potential localisation of a given activity, i.e. prognostic aspects;
- The most critical functional values and limits are those, which stem from unchangeable, partially changeable or non-renewable properties of the primary and secondary landscape structure. SEF limits must be more resilient and able to adapt to these limits and not vice versa.

These groups of limits are continuously confronted throughout the process of determination of functional values.

The prognosis values—more or less objective evaluation of expected reaction of geosystems to a certain activity—play an important role in determining the functional values and limits of all groups. This determines a logical place of the often proclaimed prognosis in a decision-making process of landscape-ecological studies: prognosis is a criterion for evaluations and propositions (Pauditšová 2014).

6.2 Formalised Decision Process in Tables

One of a very suitable formalised procedure in the process of evaluations and propositions is the systematic **decisions process in tables**, in matrices. The results of analyses were organised in the tables, as well as synthetic units of abiocomplexes ABC and landscape-ecological complexes LEC (see Tables 3.2–4.1). The tables will also be used in next steps as follows:

6.2.1 *Table of Functional Values*

The functional values of the indicators $\boldsymbol{x_n}$ should be determined by means of formalised systematic procedure, assuming clear and logical organisation of input components, which are:

- indicator of properties $\boldsymbol{x_n}$ and their values in logical scales,
- required social activities categorised according to their nature, dynamics, physical character of the CLS elements after their implementation, and according to their expected impact on the landscape, preferably in this order:

– ecostabilising activities (greenery, forests, water areas),
– agricultural activities (arable land with wide use, permanent),
– permanent crops (vineyards, orchards, gardens, gardens with cottages),
– recreational activities (cottage areas, recreational areas),
– investment activities (diverse built-up areas).

To recognise the relation between required activities ***R*** and current ***CLS elements*** is of particular importance because

- all areas of landscape-ecological complex LEC already have existing CLS elements at present which is actually the result of the given activity in the area. So, the current CLS elements provide information on **assumed influences and impact** of a particular activity ***R*** (Mander and Jongman 2000; Hreško et al. 2008). These information serve for abstract decision-making on the suitability of the LEC types for all activities, i.e. also in the case when the given activity ***does not exist yet*** on the area of evaluated LEC.

Actually, this kind of evaluation represents a strong particular stream of landscape-ecological evaluations within the frame of the well-developed process of Environment Impact Assessment (E.I.A). It is to be mentioned that the E.I.A process is not necessarily bound to any spatial planning processes. More closed to plannings is the proces of Strategic Environmental Assessment (S.E.A, Csima 2006; Balfors et al. 2010; Belčáková 2012, 2015). Based on that procedures, the evaluation process determines functional values as their suitability in a given LEC area and finally, we decide on maintaining the current use or their **change**.

- accordingly, the evaluation process includes the evaluation of all required activities ***R, also those*** which ***do not exist yet*** on the area of given LEC. So, they must be assessed according to **their expected impact** on the landscape after they possible realisation according to the experiences and knowledge about the impact of existing CLS.

So, the procedure of evaluations facilitates determining the suitability of LEC types for both the **current use** and for the **proposed use**.

The property indicators, as well as required activities, should be arranged in a table, so the functional values as described in Sect. 6.1 can be determined in this clear form. A scheme for determining functional values is shown in Table 6.1.

Note: the functional values in the table are logically correct, but they are fictive, not related to any model area. In the concrete area, they can show other values!

A specific example of simple determining of functional values is in Table 6.2 (the Ilijsky creek catchment).

The functional values for single analytical indicators from this table should be transformed to the ABC and LEC types according to their codes as displayed in combination tables (Tables 4.5 and 4.6, Sect. 4.1). The combination tables constitute the frame and outline for **decision table**.

Table 6.1 Scheme of the table for setting the functional values of the indicators

Indicator x_n	Absolute values	Functional values s_{xn}^{R} for selected activities R				
		Arable land	Forage crops	Permanent grasslands	…	Forests, greenery
x_1—slope angle	**1.**	**1.**	**3.**	**(L)**		**(L)**
	2.	***2.***	***4.***	***5.***		***(L)***
	3.	***3.***	***2.***	***4.***		***(L)***
	4.	***4.***	***1.***	***3.***		***4.***
	5.	***L***	***3.***	***1.***		***3.***
	6.	***L***	***L***	***2.***		***2.***
	7.	***0***	***0***	***L***		***1.***
x_2—soil depth and skeletality	1.	***1.***	***4.***	***(L)***		***(L)***
	2.	***2.***	***3.***	***3.***		***6.***
	3.	***3.***	***1.***	***2.***		***5.***
	4.	***4.***	***2.***	***1.***		***4.***
	5.	***L***	***5.***	***2.***		***3.***
	6.	***0***	***L***	***4.***		***2.***
	7.	***0***	***0***	***0***		***1.***
x_n—...	…					
y_1—current landscape structure	Forests	***L***	***L***	***L***		***1.***
	Shrubs	***L***	***L***	***L***		***1.***
	Marshes	***L***	***L***	***2.***		***2.***
	Water surfaces	***L***	***L***	***L***		***3.***
	Water flows	***L***	***L***	***L***		***3.***
	Agricultural land	***1.***	***1.***	***1.***		***2.***
	Residential area	***0***	***0***	***0***		***L***
	Agricultural objects	***0***	***0***	***0***		***L***
	Other objects	***0***	***0***	***0***		***0***
	Roads	***0***	***0***	***0***		***0***

The values are fictive, not related to a concrete model area

0—excluding values; *L*—limit values; (*L*)—conditionally suitable values; other numbers—suitable values in order (explanation see in Sect. 6.1)

Table 6.2 Illustration of setting the functional values of the indicators for selected activities in the Ilijský creek catchment (part)

Indicator x_n	Absolute value of indicator x_n	Code	Functional values for the selected activities R						
			Forest	Shrubs	Orchards	Grasslands	Arable land	Built-up area	Waters
			LY	*ZE*	*KU*	*TP*	*OP*	*BY*	*VO*
x_1: slope angle (accessibility) (Table 5.1)	0–1°	**1**	***(L)***	***(L)***	***(L)***	***(L)***	***1***	***1***	***1***
	1–3°	**2**	***(L)***	***(L)***	***2***	***(L)***	***2***	***2***	***(L)***
	3–7°	**3**	***(L)***	***(L)***	***2***	***3***	***3***	***3***	***(L)***
	7–12°	**4**	***3***	***3***	***1***	***2***	***L***	***L***	***0***
	12–17°	**5**	***2***	***2***	***(L)***	***1***	***L***	***L***	***0***
	17–25°	**6**	***2***	***1***	***L***	***3***	***0***	***0***	***0***
	More than 25	**7**	***2***	***2***	***0***	***L***	***0***	***0***	***0***
x_2: skeletality and depth of soils (cultivability) (Table 5.3c)	Without skelet, deep	**0**	***(L)***	***(L)***	***(L)***	***(L)***	***1***	***(L)***	***(L)***
	Without to moderate skelet, deep	**1**	***(L)***	***(L)***	***(L)***	***(L)***	***2***	***(L)***	***(L)***
	Medium skelet, deep	**2**	***2***	***4***	***1***	***2***	***3***	***(L)***	***(L)***
	Moderate skelet medium deep	**3**	***2***	***3***	***1***	***2***	***3***	***3***	***1***
	Medium skelet and deep	**4**	***1***	***2***	***2***	***1***	***4***	***2***	***1***
	Moderate skelet shallow	**5**	***3***	***1***	***3***	***2***	***L***	***2***	***1***
	Medium to skeleton, shallow	**6**	***4***	***1***	***(L)***	***3***	***L***	***1***	***1***
	Skeleton, all depths	**8**	***1***	***1***	***(L)***	***3***	***L***	***2***	***1***
	Without soils	**9**	***L***	***(L)***	***0***	***L***	***0***	***1***	***1***

(continued)

Table 6.2 (continued)

Indicator x_n	Absolute value of indicator x_n	Code	Functional values for the selected activities R						
			Forest	Shrubs	Orchards	Grasslands	Arable land	Built-up area	Waters
			LY	*ZE*	*KU*	*TP*	*OP*	*BY*	*VO*
x_3:…	…	**1**							
	…	…							
x_4:…	….	**1**							
x_n:…	…	**…**							
y_1: CLS	Forests	**1**	***1***	***L***	***L***	***L***	***L***	***L***	***L***
	Shrubs	**2**	***2***	***1***	***(L)***	***(L)***	***(L)***	***(L)***	***(L)***
	Permanent cultures	**3**	***L***	***L***	***1***	***3***	***(L)***	***(L)***	***(L)***
	Pastures	**4**	***2***	***2***	***2***	***1***	***4***	***2***	***2***
	Meadows	**5**	***2***	***2***	***2***	***1***	***3***	***3***	***2***
	Forage crops	**6**	***2***	***2***	***2***	***2***	***2***	***4***	***2***
	Arable land	**7**	***2***	***2***	***2***	***3***	***1***	***(L)***	***(L)***
	Residential houses	**8**	***L***	***3***	***L***	***L***	***L***	***1***	***L***
	Technical objects	**9**	***L***	***3***	***L***	***L***	***L***	***1***	***L***
	Waters	**10**	***L***	***L***	***L***	***L***	***L***	***L***	***1***
	Bare surfaces	***n***	***L***	***2***	***L***	***L***	***L***	***1***	***1***

Table 6.3 Scheme of the functional parts (zones) of the decision table—the logic of the decision in table

*Zone of **analyses and syntheses** of primary landscape structure* *Localisation criteria*				*Zone of **evaluation***										*Zone of **analyses** of the current landscape structure CLS* *Selective criteria*						
				***Functional values** of ABC for activities R - s^{R}_{ABC} according to localisation criteria*								***Negative choice***								
Primary landscape sructure - **types of ABC ($x_1, x_2, \ldots, x_n$)**				**Funtional values s^{R}_{ABC}** (transformed from Tab.40)								**Results**		**CLS – current landscape structure - y_n**						
x1 – slope angle	x2 – soil skeletality and depht	xn - ?	ABC type ode	Forest	Shrubs	Orchards	Grasslan	Forage	Arable	Built-up	Waters	Activities without limits	sequence	Code of the element of CLS						
														1	2	3	4	5	6	7
1st code value	2nd code value		nn...n	LE	ZE	KU	TP	KR	OP	BY	VO			Forest	Shrubs	Orchard	Grassland	Arable land	Built-up area	Water
*Values of the **ABC** indicators*				***Functional values** s^{R}_{xi} according to localisation*								***List of activities***		*Zone of **LEC synthesis*** *Zone of **propositions*** • *Application of the CLS preferences* • *Alternative propositions according to the selective criteria* • *Proposition variants*						

6.2.2 Decision Table

Decision table is the commonly used formal method of assigning functional values as well as other functional values to actual LEC types in the LANDEP (continually developed as in Ružička and Miklós 1982, 1990; Miklós et al. 1986; Miklós and Špinerová 2011). The scheme of a decision table is shown in Table 6.3.

Description of functional parts (zones) of decision table:

A decision table is divided into functional parts with clearly defined functions as follows:

- Zone of analyses and syntheses of the primary landscape structure—abiotic complex ABC.

This part of the table is the transfer of the left-hand section of the combination Table 4.2 and contains systematically arranged values of ABC indicators **ABC** ($x_1, x_2, \ldots, x_n$) in the rows of the table. The column under mark **ABC type code** is the consequent combination of the single indicators, i.e. each row of this column encompasses one **type of ABC** identified by a numerical code formed by a combination of codes of the indicator values of analytical x_n. These values form a base for deriving localisation criteria to determine functional indicator values.

- Zone of evaluations—determining functional values.

This part of the table supports the systematic initial confrontation of the indicators with ABC complexes and required activities. This part is entered cardinal **functional values** s_x^R of single x_n—namely zero values—**0**, limit—***L***, conditionally suitable—(***L***) and other values of suitability degree—consecutively for all indicator in **ABC** type and each activity ***R***.

- Zone evaluations—negative selection.

It is a transitional zone between evaluations and propositions;
This section of the table holds the result of **limitations**, the list of activities without limit values. The next column can contain unlimited activities, which can be arranged according to the degree of suitability to sequence. Then, this **negative selection** can be considered as a primary 'proposition' of limitation of non-suitable activities.

- Zone of analyses of the current landscape structure CLS.

This section of the table is identical with the header of the right-hand section of the combination Table 4.5. In the header, it contains ordered **CLS** elements in the functional part.

- Zone of LEC syntheses.

A part of this section of the table is identical with the header of the right-hand section of the combination Table 4.6. The cells presents the combinations of the CLS elements (columns) with ABC types (rows), so the cells contain the combination of the codes of **LEC** [**ABC** $\{x_1, \ldots, x_n\}$, **CLS** $\{y_1\}$] as generated in the Table 4.5.

- Zone of propositions.

It occupies the same space as the zone of synthesis of LEC types, which is logical as this phase of propositions is conducted by applying selective criteria—based on CLS elements. This section determines a modification of the initial proposals gained from a negative selection, so we assess the suitability of activities, which were not limited by ABC types, based on a CLS element occurring in a given ABC.

Projection of Chosen Functional Values from a Decision Table Onto a Map

The most common output from the LANDEP method is the projection of the limit values onto maps. As the limit values were determined for each ABC type, they can be transferred from the decision table into maps of ABC types according to the codes of types of abiotic complexes. This process is very common in different studies, in particular with regards to the optimisation of the agricultural use of the landscape (Beek and Bennema 1972; Džatko et al. 1976; McCrae and Burnham 1981; Young 1986; Beek et al. 1987; Schindler et al. 2009).

Schematic procedure of a transfer of these values is presented in two steps as follows:

– transfer the limit values of single indicators $\boldsymbol{x_n}$ for activities **R** from Table 6.1 to the types of ABC to corresponding rows according to the values of ABC and corresponding columns of activities under the header **limit and zero values** $\boldsymbol{s^R_{ABC}}$ into the Table 6.4
– listing the unlimited activities for ABC types in the column **Results**.

Table 6.4 Scheme of the transfer of limit values to the decision table

Primary landscape sructure - **types of ABC ($x_1, x_2, \ldots, x_n$)**				**Limit and zero values s^R_{ABC}** (transformed from Tab.40)								**Results**		**CLS – current landscape structure - y_n**						
x_1 – slope angle	x_2 – soil depht	x_n - ?	ABC type ode	Forest	Shrubs	Orchards	Grassland	Forage	Arable	Built-up	Waters	Activities without limits	Sequence of suitability	Code of the element of CLS						
														1	2	3	4	5	6	7
1st code value	2nd code value	…	nn…n value	LE	ZE	KU	TP	KR	OP	BY	VO			Forest	Shrubs	Orchard	Grassland	Arable land	Built-up area	Water
1 *(0-1°)*	1 *(deep)*		11…	(L) (L)	(L) (L)	(L) (L)	(L) (L)					KR,OP	OP,KR							
	2 *(medium)*		12…	(L)	(L)	(L)	(L)					KR,OP,BY,V	KR,OP,BY							
	3 *(shallow)*		13…	(L)	(L)	(L) (L)	(L)	L	L			BY, VO	BY,VO							
	4 *(without soils)*		14…	(L) L	(L) L	(L) 0	(L) L	0	0			BY,VO	BY,VO							
2 *(1-3°)*	1 *(deep)*		21…	(L) (L)	(L) (L)	(L)	(L) (L)					KR,OP	OP,KR							
	2 *(medium)*		22…	(L)	(L)		(L)					KU,KR,OP,B	K,O, K,B							
	3 *(shallow)*		23…	(L)	(L)	(L)	(L)	L	L			BY	BY							
	4 *(without soils)*		24…	(L) L	(L) L	0	(L) L	0	0			BY	BY							
3 *(3-7°)*	1 *(deep)*		31…	(L) (L)	(L) (L)	(L)	(L)					KR,OP	KR,OP							
	2 *(medium)*		32…	(L)	(L)							K,T,K,O,B	K,K,O,TP, BY							
	3 *(shallow)*		33…	(L)	(L)	(L)		L	L			TP,BY	TP,BY							
	4 *(without soils)*		34…	(L) L	(L) L	0	L	0	0			BY	BY							
4 *(7-12°)*	1 *(deep)*		41…	(L)	(L)	(L)	(L)		L			KR								
	2 *(medium)*		42…						L			LE,Z,K,T,KR								
	3 *(shallow)*		43…			L		L	L L			LE,ZE,TP								
	4 *(without soils)*		44…	L	L	0	L	0	L 0			-	-							
5 *(12-17°)*	1 *(deep)*		51…	(L)	(L)	(L) (L)	(L)	L	L				TP,ZE,LE							
	2 *(medium)*		52…			(L)		L	L			LE,ZE, TP	LE,ZE,TP							
	3 *(shallow)*		53…			(L) L		L L	L L			LE,ZE,TP	ZE,LE,TP							
	4 *(without soils)*		54…	L	L	(L) 0	L	L 0	L 0			-	-							
6 *(17-25°)*	1 *(deep)*		61…	(L)	(L)	L (L)	L (L)	0	0				LE.ZE							
	2 *(medium)*		62…			L	L	0	0				LE,ZE							
	3 *(shallow)*		63…			L L	L	0 L	0 L				ZE,LE							
	4 *(without soils)*		64…	L	(L)	L 0	L L	0 0	0 0				-							
7 *(>25°)*	1 *(deep)*		71…	(L)	(L)	0 (L)	L (L)	0	0	0 (L)	0 (L)		LE,ZE							
	2 *(medium)*		72…			0	L	0	0	0	0		LE,ZE							
	3 *(shallow)*		73…			0 L	L	0 L	0 L	0	0		ZE,LE							
	4 *(without soils)*		74…	L	L	0 0	L L	0 0	0 0	0	0		-							

This procedure is demonstrated also on the example of the Ilijský stream catchment, where Fig. 6.1 depicts the projection of simple land-use limits onto a map (Špinerová 2010).

Determination of the limits of land use and their projection onto a map is currently
the most exploitable output
of the LANDEP methodology

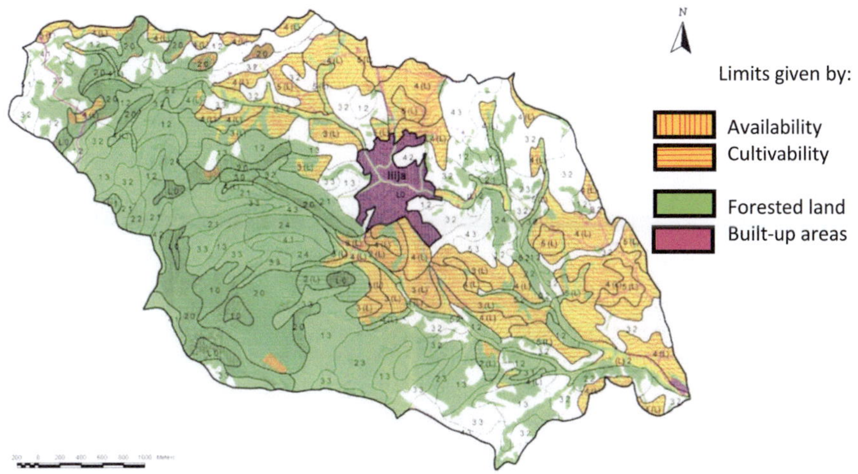

Fig. 6.1 Ilustration of the projection of chosen limitations of the agricultural use of the areas of LEC types (Limits of availability are determined by slope angle, limits of cultivability by depth and skeletality of soils, real usability by CLS on forested land and built-up areas.) Functional values and limitations may occur in combinations each with others given by numbers in LEC type areas (Table 6.3). First number—functional value for availability, Second—for cultivability

6.2.3 Weight Coefficients and Total Functional Suitability of LEC for Activities **R**

The importance of different indicators of properties x_n vary for different activities. E.g. the indicators of slope inclination, depth and skeletality of soils are significant for arable land. For the construction of houses is more important orientation towards the cardinal points. For nature conservation is most important the species composition of vegetation, which is irrelevant for housing construction and farming. A situation may arise in which some indicators exhibit high suitability, but other indicators have limiting character. Probably even very high functional suitability of macroclimate cannot compensate for, e.g. unfavourable value of high slope inclination, i.e. in this case, the macroclimate is less important and slope inclination is a crucial parameter, and gentle slopes are suitable for vineyards, but northern exposure makes such activity impossible, but it is suitable for localisation of a ski slope. Each project may deliver many similar considerations. This aspect of indicators is in the LANDEP methodology formulated as weighting coefficients.

Weighting coefficients v_{xn}^{R} express the importance (significance) of a given property x_n in comparison with other properties for a given activity R. They play an important role in determining total functional suitability. In other words, weighting coefficients in evaluation equation balance the fact that not each indicator is equally important for individual activities.

Table 6.5 Example of the assessment of the weight coefficients of interpreted indicators j_n for selected activities on model territory East-Slovakian lowland

Example for ***R* – *Arable land***

	a	*c*	*p*	*w*	*e*	*i*	*n*	*v*	Σ
a		1	1	0,5	1	1	1	0	**5,5**
c	0		1	1	1	1	1	0	**5**
p	0	0		1	0,5	1	1	0	**3,5**
w	0,5	1	1		1	1	1	0	**5,5**
e	0	0	0,5	0		1	1	0	**2,5**
i	0	0	0	0	0		0	0	**0**
n	0	0	0	0	0	1		0	**1**
v	1	1	1	1	1	1	1		**7**

Example for ***R* – *Residental buildings***

	a	*c*	*r*	*w*	*e*	*i*	*n*	*v*	Σ
a		1	1	0	1	1	1	0	**5,5**
c	0		1	0	1	1	1	0	**4,5**
p	0	0		0	0	0,5	0,5	0	**1,5**
w	1	1	1		1	1	1	0	**6,5**
e	0	0	1	0		1	1	0	**3,5**
i	0	0	0,5	0	0		0,5	0	**1,5**
n	0	0	0,5	0	0	0,5		0	**1,5**
v	1	1	1	1	1	1	1		**7,5**

The assessment process is as follows:
First step: assessment of partial weight of pairs of indicators
If indicator j_i against indicator, j_k is assessed as
—more important, partial weight coefficient = 1
—equally important, partial weight coefficient = 0,5
—less important, partial weight coefficient = 0
Second step: calculation of complex weight coefficient
The complex weight coefficient v_{xn}^{R} = the sum of partial weight coefficients in rows of the tables
Assessed indicators: *a*—accessibility (slope angle); *c*—cultivability (depth and skeletality of soils); *p*—ploughibility (soil texture); *w*—waterlogging (complex interpretation of LEC; *e*—erodibility by water erosion (complex interpretation), *i*—erodibility by wind erosion (complex interpretation); *n*—nutritiousness (complex interpretation); *v*—preciousness of the vegetation (complex interpretation)

They can be determined by employing various methods. The LANDEP methodology juxtaposes the importance of pairs of indicators in a table (simplified Fuller's triangle method). The process of determining weighting coefficients for selected activities is presented in Table 6.5 (Miklós et al. 1986b).

The Overall LEC *Suitability* W_{LEC}^{R}

The overall LEC suitability for activities *R* is its suitability based on both,

- the indicators of the properties x_n and
- weight coefficients v_{xn}^{R}.

A formal systematic expression of the evaluation process draws on the fact that all properties simultaneously affect the conditions of performing selected activities in an examined territory, therefore, the total suitability of a given ecosystem for a given activity ***R*** can be expressed by a conceptual model as follows:

$$W_{\mathrm{LEC}}^{R} = \sum_{x=1}^{n} s_{xi}^{R} \cdot v_{xi}^{R}$$

where

W total suitability of LEC types for activity ***R***,

s^R_{xi} functional value of properties of *i*th indicator ***x*** for activity ***R*** in a type **LEC;**
v^R_{xi} weighting coefficient of the ***ith*** indicator ***x*** for activity ***R.***

The result may be expressed as the proportional suitability of a particular LEC type for the selected social activities against the maximum suitability (Miklós et al. 1986). The whole evaluation process can be then transferred to computer to be processed by GIS.

It is crucial to employ limits in the evaluation formula to determine the total suitability of a given territory correctly: in case of the occurrence of even one limit value for any indicator in a given **LEC**, this **LEC** is considered unsuitable for a given activity, even if all remaining indicators exhibit high functional values (Miklós et al. 1986). The comprehensive limitation is then a negative selection of possible activities for a given area.

Methodological Overview of Evaluations

It should be stressed that in many cases, decisions regarding suitability of land use are reduced to defining of limit values, i.e. determining interfaces between suitable and unsuitable activities in a given area. The result of such simplified evaluation is so-called **negative selection**, which specifies which activities are not possible or limited in a given area. Although this result is not considered optimal from the LANDEP point of view, we concede that if the cardinal values are determined correctly by eliminating unsuitable activities, we can already prevent the consequences of unsuitable activities. This strategy is sometimes referred to as a **selection of the lesser evil**.

Taking into account the current situation of development of the society and actual status of spatial planning processes, it is necessary to critically express that users and owners of land as well as developers do not like to be told what activity is and is not suitable in a given territory, and they want to determine it by themselves based on their economic interests. Therefore, the success of a landscape-ecological plans is already when at least the limits and restrictions defined in the landscape-ecological plan are accepted.

The situation could be remedied by new act on landscape planning, which would represent an integrated *spatial projection* of all interests of *protection of nature, landscape and environment*, will be an integrated *complex of binding regulations* for environmental care, for administrative proceedings as well as *integrated document* for all types of spatial planning processes.

References

Balfors B, Mörtberg U, Geneletti D (2010) Landscape ecology for SEA: lessons learned. IAIA10 conference proceedings. The role of impact assessment in transitioning to the green economy 30th annual meeting of the international association for impact assessment 6–11 April 2010. International Conference Centre, Geneva

Barsch H, Saupe G et al (1993) Zur Integration landschaftsoekologischer und sociooekologischer Daten in gebietliche Planungen. Potsdamer Geographische Forschungen, Band 4. Universität Potsdam

Bastian O, Schreiber KF (1994) Analyse und ökologische Bewertung der Landschaft. Gustav Fischer Verlag Jena, Stuttgart

Bastian O, Kronert R, Lipsky Z (2006) Landscape diagnosis on different space and time scales—a challenge for landscape planning. Landscape Ecol 21:359–374

Beek KJ, Bennema J (1972) Land evaluation for agricultural land use planning. An ecological methodology. Department of Soil Science and Geology, Agricultural University, Wageningen

Beek KJ, Burrough PA, McCormack DE (ed) (1987) Quantitative land evaluation procedures. ITC Publication 6, Enschede

Belčáková I (2012) Landscape planning framework in the environmental assessment—linkages and mutual benefits. Ecology 31(1):1–11

Belčáková I (2015) Hodnotenie vplyvov na krajinu v plánovacom procese. Bratislava, Veda

Boltižiar M (2011) Inventory and classification of historical structures of the agricultural landscape in Slovakia. Ekológia 30(2):157–170

Csima P (2006) Tájvédelmi szabályozás a településrendezési tervekben. In: Csorba P, Fazekas I (eds) Tájkutatás – Tájökológia. Meridián Alapítvány, Debrecen

Csorba P, Szabó S (2012) The application of landscape indices in landscape ecology. In: Tiefenbacher J (ed) Perspectives on nature conservation—patterns, pressures and prospects, pp 121–140. InTech, Available from http://www.intechopen.com/books/perspectives-on-natureconservation-patterns-pressures-and-prospects/the-application-of-landscape-indices-in-landscape-ecology

Diviaková A (2011) Biotické komplexy pre environmentálny manažment. VKÚ, a. s., Harmanec

Džatko M et al (1976) Charakteristika bonitovaných pôdno-ekologických jednotike SSR. MPVž, Bratislava

Hreško J, Mederly P, Petrovič F (2003) Landscape ecological research with support of GIS tools in preparation of landscape-ecological plan (model area of the Považská Bystrica city). Ekológia 22(2):195–212

Hreško J, Bugár G, Petrovič F, Mederly P (2008) Endangering of abiotic complex as a basis for evaluation in the landscape-ecological plan of the Trenčín catastral area (Slovakia). Gospodarka surowcami mineralnymi. 24(2):156–158

Hrnčiarová T (1999) Krajinnoekologické plánovanie pomocou metodiky LANDEP a metodiky EÚK. Geografický časopis 51(4):399–413

Hrnčiarová T, Miklós L et al (1997) Ekologická únosnosť: metodika a aplikácia na 3 benefičných územiach. I.-IV. Bratislava, Ústav krajinnej ekológie SAV a Ministerstvo životného prostredia SR

Jacobs S et al (2016) A new valuation school: integrating diverse values of nature in resource and land use decisions. Ecosyst Serv 22(12):213–220

Jones KB, Zurlini G, Kienast F, Petrosillo I, Edwards T, Wade TG, B-l Li, Zaccarelli N (2013) Informing landscape planning and design for sustaining ecosystem services from existing spatial patterns and knowledge. Landscape Ecol 28:1175–1192

Kozová M, Bedrna Z (eds) (2003) Krajinnoekologické metódy v regionálnom environmentálnom hodnotení. Univerzita Komenského v Bratislave

Kozová M, Miklós L, Ružička M, Ružičková H, Terek J et al (1986) Ekologické predpoklady pre rozvoj poľnohospodárskej výroby SŠM Somotor (VSN) v mierke 1:10 000. Ekologická optimalizácia využívania VSN. ÚEBE SAV Bratislava, Slovosivo. III. diel, pp 353–410

Kozová M, Pauditsová E, Finka M (eds) (2010) Krajinné plánovanie. STÚ Bratislava, UK Bratislava, UKF Nitra, SPU Nitra

Lóczy D (2002) Landscape evaluation, land evaluation. Dialóg Campus, Pécs

Löw J et al (1995) Rukověť projektanta místního ÚSES. Agroprojekt, Brno

Mander Ü, Jongman R (ed) (2000) Consequences of land use changes. In: Advances in ecological sciences series 5, WIT Press, pp. 11–38. ISSN:1369-8273, Southampton, Boston, USA

McCrae SG, Burnham CP (1981) Land evaluation. Clarendon, Oxford

Mederly P et al (2015) Integrated valuation of ecosystem service as a for landscape management and decision making at regional and local level (case study Trnava, Slovakia). In: Ecosystem service for nature, people and prosperity, ecosystem services partnership

Miklós L et al (1986) Ekologický plán VSN. Súbor grafických výstupov. Ekologická optimalizácia využívania VSN. IV. diel. ÚEBE SAV Bratislava, Slovosivo

Miklós L, Špinerová A (2011) Krajinno-ekologické plánovanie LANDEP (The Landscape-ecological Planning LANDEP). Harmanec, VKU

Miklós L, Kozová M, Ružička M et al (1986a) Ekologický plán využívania Východoslovenskej nížiny v mierke 1:25 000. Ekologická optimalizácia využívania VSN. III. diel. ÚEBE SAV Bratislava, Slovosivo, pp 5–312

Miklós L, Miklisová D, Reháková Z (1986b) Systematisation and Automatisation of Decision-Making Process in LANDEP method. Bratislava, Ecology (CSSR), 5(2):203–232

Mizgajski A, Stępniewska M (2012) Ecosystem services assessment for Poland—challenges and possible solutions. Ekonomia i Środowisko 2(42):54–73

Múdry P et al (1995) Uplatnenie princípov ekologickej únosnosti pri stanovení ekologických regulatívov územného rozvoja. IV. časť. Záverečná správa projektu, EKOTRUST, Banská Štiavnica

Pauditšová E (2014) Hodnotenie vplyvov na krajinu v procese posudzovania vplyvov na životné prostredie. Acta Environmentalica Universitatis Comenianae 22(1):72–96

Ružička M, Miklós L (1982) Metodické poznatky ekologického hodnotenia územia pre zónu a sídelný útvar (na príklade Rimavskej Soboty). VEDA, Bratislava, Acta Ecologica, vol 9, Issue 26, 74 pp

Ružička M, Miklós L (1990) Basic premises and methods in landscape-ecological planning and optimisation. In: Zonnenveld IS, Forman RTT (eds) Changing landscapes: an ecological perspectives. Springer, New York, pp 233–260

Schindler S, Wrbka T, Peterseil J, Pollheimer M, Schmitzberger I (2009) Impact of the Austrian agri-environmental scheme on diversity of landscapes; plants and birds. In: Breuste J, Kozová M, Finka M (eds) European landscapes in transformation: challenges for Landscape ecology and management (PROCEEDINGS of the European IALE conference), p 193

Sklenička P (2003) Základy krajinného plánování. Vyd. Naděžda Skleničková, Praha

Špinerová A (2010) Krajinno-ekologické limity poľnohospodárskeho využitia Ilíjskeho potoka. VKÚ, a.s., Harmanec

Špinerová A (2015) Štruktúra krajiny ako regulátor dynamiky pohybu vody a materiálu. Vydavateľstvo TU vo Zvolene, Zvolen

Špulerová J, Dobrovodská M, Izakovičová Z, Kenderessy P, Petrovič F, Štefunková D (2013) Developing a strategy for the protection of traditional agricultural landscapes based on a complex landscape-ecological evaluation (the case of a mountain landscape in Slovakia). Moravian Geogr Rep 21(4):15–26

Young A (1986) Land evaluation and diagnosis and design: towards a reconciliation of procedures. Soil Surv Land Eval 5:61–76

Chapter 7
Landscape-ecological Propositions in LANDEP—Proposal of Optimal Use of Areas

Abstract The objective of this stage is to formulate a proposal of an ecologically optimal utilisation of each type of landscape-ecological complexes. This proposition should provide a frame for the ecologically optimum organisation of the territory. The propositions have several stages using different applications of criteria. The **negative selection** is based on the application of limiting values which determine the unsuitable activities, thus, on the other side, also those, which are suitable for a given area. The primary **alternative proposition** results in the determination of few (one to three) most suitable activities—alternatives—for each type of landscape-ecological complexes. Secondary **variant proposition** assigns an ecologically most suitable use (a function) for each type of landscape-ecological complexes by selection from the alternatives yielded by the primary proposition. The result of secondary proposition is so-called **functional typisation** of territory. The functional typisation of territory involves the designation of:

- suitable areas for different agricultural activities;
- suitable areas of green infrastructure, water areas, wetlands and forests;
- suitable areas for recreational use and areas for residential buildings;
- suitable areas for industrial and technical objects.

This result mainly serves as a base for physical/spatial/territorial plans of municipalities and their functional zones. Tertiary proposition—**functional regionalisation** of territory—aims to determine the most ecologically appropriate function of larger territorial units (subregions, regions) in terms of their broader territorial (choric) relations. The result should serve as the basis for physical/spatial/territorial plans of a region. Special and detailed proposals may have very different characters. The most common examples are the projects of territorial system of ecological stability TSES that are specifically defined in the legislation. Other partial outputs are: suitability of territory for hydro-melioration, for establishment of water areas, ponds, solving the problems of erosion and accumulation, proposal of landscape greenery, proposal of recreational utilisation of the landscape, problems related to the conflict of interests, spatial projection of the limits of territory for selected activities, problems of flooding, etc. The chapter is illustrated by numerous tables and maps.

L. Miklós and A. Špinerová, *Landscape-ecological Planning LANDEP*,
https://doi.org/10.1007/978-3-319-94021-2_7

Keywords Negative selection · Alternative proposition · Functional typisation
Functional regionalisation

The basic objective of the propositions in the LANDEP methodology is to formulate a proposal of an ecologically optimal utilisation of each LEC type. This proposition should provide a frame of the ecologically optimum organisation of the territory. This frame is then modified by further considerations.

The proposals are based on an evaluation procedure, which determines

- the suitability of a current use of a LEC type;
- the total suitability of a LEC type for individual activities;
- other criteria of optimal use, which cannot be always parameterised, e.g. requirements to maximise intensification of land use, requirements for protection and development of the landscape, etc.

The propositions have several stages using different applications of criteria:

7.1 Negative Selection

This is a transitional step between evaluations and propositions because it represents only the transfer of results of comprehensive limitation onto a map. A complex of limits in respective areas of LEC types based on of '**negative selection**' then determines which activities are unsuitable, but also those which are possible and suitable for a given area. The basic process of such approach—including tables and figures—is described in Sect. 6.2. However, this approach can be developed in different directions.

7.1.1 Negative Selection by Comprehensive Limitation

Determining unsuitable and possible/suitable activities itself can be regarded as a certain kind of proposal. This method can be successfully applied to territories that include numerous limit indicator values and LEC types are suitable only for a few or just one activity. Such 'stenoec' types can be found particularly in mountainous regions with poor soils and high proportion of forests. Conversely, in lowlands, most LEC types are suitable for a wide range of activities, they are universal and it is important to employ other methods of proposals here.

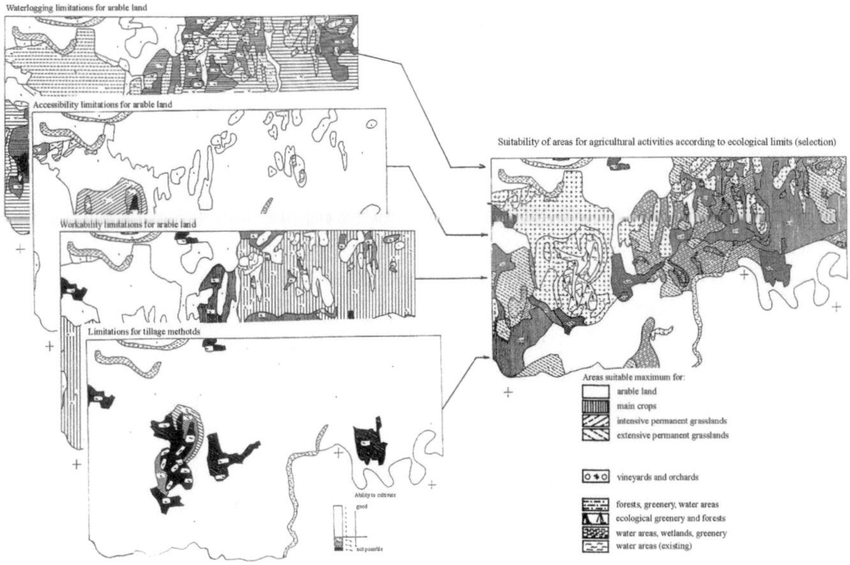

Fig. 7.1 Example of the proposal based on negative selection according to partial limits for arable land (East Slovakian Lowland)

If we manage to construct a scale of limits and a succession of limited activities correctly, based on negative selection, it will be possible to draw up a proposal for suitable use of land. As an example, we present a proposal of the project of ecological optimisation of use of the Eastern Slovak lowland. In this case, the limits were set from the perspective of agricultural activities, whereas gradually worse conditions limited the range of activities with gradually diminishing demands on conditions as: arable land → forage crops → intensive grasslands → moderately intensive grasslands → extensive grasslands. Conversely, it was possible to propose only extensive grasslands for the worst conditions and arable land only after the conditions improved. These basic activities were supplemented by permanent crops, forest and landscape greenery (Fig. 7.1).

7.1.2 Application of Preference Limits, Restrictions and Proposals

To be able to decide on the best use in highly differentiated territory with a big number of types of abiocomplexes ABC and diverse current landscape structure CLS needs proper simplifying methods. The most handy approach is the lowering of the number of ABC/CLS combinations. So, as an early step of proposition process, we try to eliminate from the decision making process such ABC/CLS combinations which are unfeasible. However, if such combinations occure they need to be changed. These decisions can be referred to as preference decisions. Basically, we recognise two types of such decisions:

(a) The exclusion of defined unreal combinations of ABC and CLS types from decision-making:

- ABC types without soil, but mapped with CLS arable land—it is a nonsense, as a result of a mistake, made during mapping of the soils or the CLS;
- ABC types without soil but mapped as forests, greenery and grasslands—improbable combinations. These uses cannot exist without soil, and it is necessary to preferentially examine a combination;
- ABC types with slope inclinations over 25° intended for agricultural activities—this is also an extreme combination and it is necessary to examine the probability of a mistake and change the combination;
- ABC types with slope inclinations over 17° intended for housing—improbable combination, which is necessary to examine and change.

(b) Preference of current land use as unambiguous proposal for further use:

This emphasises the maintaining of a current land use regardless of the limits of abiotic complexes, which may occur in a given area. Such land uses are desirable from the ecological point of view, represent permanent use of land, or their change would affect the landscape to such an extent that any new uses could not compensate. This approach includes the following preferences:

- Keeping the forests and other green infrastructure in landscape—these should never be proposed for elimination or change regardless of the limits or other values of abiocomplexes;
- Water areas will not be proposed for elimination because water areas are indispensable in the landscape and there is no guarantee that the bottom of an eliminated water body would provide an adequate compensation if the areas were not used as water areas;
- Built-up areas will be proposed for a different use only in exceptional cases, with the consideration that urban and housing development should be promoted more intensively in the areas already used in that capacity, and no new areas should be seized for that purpose. On the other hand, the use of built-up areas for other purposes—considering the physical character of an area after demolishing buildings—is really questionable;
- Similarly, permanent crops—as orchards, vineyards, gardens—should not be proposed for elimination because their planting and growing require a long time and their elimination would inflict substantial material damage just after a short time. In addition, permanent crops also have a considerable ecostabilising importance.

The overview of these cases of preferences is given in two steps—as unreal combinations of ABC and CLS types and preference of current land use are shown in the following graphics:

	Impossible or unreal combinations of the values of individual ABC elements, e.g.—ABC over 25° with deep soil Impossible and unreal combinations of the properties of ABC elements with CLS elements, e.g. – ABC without soil/CLS—arable land – ABC over 25°/CLS—any agricultural crops – ABC over 7°/CLS of water
	Improbable combination of values of individual ABC elements, e.g. – ABC over 17° with deep soil, arable soil and – Improbable combination of values of individual ABC elements with CLS elements, e.g. – ABC without soil/CLS forests, greenery, grasslands – ABC over 17°/CLS built-up areas
	Probable preference CLS—forests and greenery will not be eliminated!
	Probable preference CLS—water areas will not be eliminated!
	Probable preference CLS—built-up areas will not be eliminated!
	Probable preference CLS—permanent crops (orchards, vineyards, gardens) will not be eliminated!

The graphics serve also as the colour legend key to the Table 7.1.

Table 7.1 shows actually the continuation decision process on the decisions table whereby the basic frame of the table and the results of the foregoing steps are maintained.

The organisation of these preferences into a table would eradicate a considerable number of combinations, which do not deserve any further consideration and other evaluations and proposals can be addressed instead. Therefore, this procedure can also be considered as a negative selection.

The procedure of preferential decisions and the application of preference limits are shown in Table 7.1 by means of coloured, highlighting those ABC/CLS combination in the column **CLS—current landscape structure—y_n** which:

- are mapped by mistake, they probably do not exist, therefore do not require proposals. Nevertheless, it is necessary to identify a potential error in the combination. These are in the table highlighted in red and orange;

Table 7.1 The completed process on decision table according to the limits, preferences and values of LEC types

Primary landscape sructure—types of ABC ($x_1, x_2, \ldots, x_n$)				Limit and zero values s^R_{ABC} (transformed from Table 6.3)								Results		CLS—current landscape structure—y_n						
x_1—slope angle	x_2—soil depht	x_n —?	ABC type ode	Forest	Shrubs	Orchards	Grassland	Forage land	Arable	Built-up	Waters	Activities without limits	Sequence of suitability	Code of the element of CLS						
														1	2	3	4	5	6	7
1st code value	2nd code value	…	nn…n value	LE	ZE	KU	TP	KR	OP	BY	VO			Forest	Shrubs	Orchard	Grassland	Arable land	Built-up area	Water
1 (0–1°)	1 *(deep)*		11…	(L) (L)	(L) (L)	(L) (L)	(L) (L)					KR, OP	OP, KR	M = LE	M = ZE	M = KU	M = OP	P = OP	M = BY	M = VO
	2 *(medium)*		12…	(L)	(L)	(L)	(L)					KR, OP, BY, V	KR, OP, BY	M = LE	M = ZE	M = KU	M = KR	P = OP	P = BY	P = VO
	3 *(shallow)*		13…	(L)	(L)	(L) (L)	(L)	L	L			BY, VO	BY, VO	M = LE	M = ZE	M = KU	M = TP	N = BY	P = BY	P = VO
	4 *(without soils)*		14…	(L) L	(L) L	(L) 0	(L) L	0	0			BY, VO	BY, VO	M = LE	M = ZE		N = BY		P = BY	P = VO
2 (1–3°)	1 *(deep)*		21…	(L) (L)	(L) (L)	(L)	(L) (L)					KR, OP	OP, KR	M = LE	M = ZE	M = KU	M = OP	P = OP	M = BY	M = VO
	2 *(medium)*		22…	(L)	(L)		(L)					KU, KR, OP, B	K, O, K,B	M = LE	M = ZE	P = KU	M = KR	P = OP	P = BY	M = VO
	3 *(shallow)*		23…	(L)	(L)	(L)	(L)	L	L			BY	BY	M = LE	M = ZE	M = KU	M = BY	N = BY	P = BY	M = VO
	4 *(without soils)*		24…	(L) L	(L) L	0	(L) L	0	0			BY	BY	M = LE	M = ZE		N = BY		P = BY	M = VO
3 (3–7°)	1 *(deep)*		31…	(L) (L)	(L) (L)	(L)	(L)					KR, OP	KR, OP	M = LE	M = ZE	M = KU	M = KR	P = OP	M = BY	M = VO
	2 *(medium)*		32…	(L)	(L)							K, T, K, O, B	K,K, O, TP, BY	M = LE	M = ZE	P = KU	P = KR	P = OP	P = BY	M = VO
	3 *(shallow)*		33…	(L)	(L)	(L)		L	L			TP, BY	TP, BY	M = LE	M = ZE	M = KU	P = BY	N = TP	P = BY	M = VO
	4 *(without soils)*		34…	(L) L	(L) L	0	L	0	0			BY	BY	M = LE	M = ZE		N = BY		P = BY	M = VO
4 (7–12°)	1 *(deep)*		41…	(L)	(L)	(L)	(L)		L			KR		M = LE	M = ZE	P = KU	M = KR	N = KR	M = BY	
	2 *(medium)*		42…						L			LE, Z, K, T, KR		P = LE	P = ZE	P = KU	P = TP	N = KR	M = BY	
	3 *(shallow)*		43…			L		L	LL			LE, ZE, TP		P = LE	P = ZE	P = KU	P = TP	N = TP	M = BY	
	4 *(without soils)*		44…	L	L	0	L	0	L0			–	–	P = LE	P = ZE	N = ZE	N = ZE		M = BY	
5 (12–17°)	1 *(deep)*		51…	(L)	(L)	(L) (L)	(L)	L	L				TP, ZE, LE	M = LE	M = ZE	M = KU	M = TP	N = TP	N = BY	
	2 *(medium)*		52…			(L)		L	L			LE,ZE, TP	LE, ZE, TP	P = LE	P = ZE	M = KU	P = TP	N = TP	N = BY	
	3 *(shallow)*		53…			(L) L		LL	LL			LE,ZE,TP	ZE, LE, TP	P = LE	P = ZE	M = KU	P = TP	N%TP	N = BY	
	4 *(without soils)*		54…	L	L	(L) 0	L	L0	L0			–	–	N = LE	N = ZE	N = ZE	N = ZE		N = BY	
6 (17–25°)	1 *(deep)*		61…	(L)	(L)	L (L)	L (L)	0	0				LE.ZE	M = LE	M = ZE	N = KU	N = LE	N = ZE	N = BY	
	2 *(medium)*		62…			L	L	0	0				LE, ZE	P = LE	P = ZE	N = KU	N = TP	N = ZE, TP	N = BY	
	3 *(shallow)*		63…			LL	L	0L	0L				ZE, LE	P = LE	P = ZE	N = KU	N = ZE	N = ZE	N = BY	
	4 *(without soils)*		64…	L	(L)	L0	LL	00	0 0				–	N = LE	P = ZE	N = ZE	N = ZE		N = BY	
7 (>25°)	1 *(deep)*		71…	(L)	(L)	0 (L)	L (L)	0	0	0 (L)	0 (L)		LE, ZE	P = LE	P = ZE	N = LE	N = LE			
	2 *(medium)*		72…			0	L	0	0	0	0		LE, ZE	P = LE	P = ZE	N = LE	NE = LE			
	3 *(shallow)*		73…			0L	L	0L	0L	0	0		ZE, LE	P = LE	P = ZE	N = ZE	NE = ZE			
	4 *(without soils)*		74…	L	L	00	LL	00	00	0	0		–	P = LE	P = ZE	N = ZE	NE = ZE			

- where the current use of the ABC types should be preserved regardless of ABC/CLS combination (highlighted in any other colour).

Opposite to the eliminated combinations, Table 7.1 shows those combinations of ABC/CLS which need considerations on their suitability for current use in terms if to keep current CLS or change. Those combinations in the column **CLS—current landscape structure—y_n** are not coloured. They represent actually the LEC with arable land and grassland. This means, in the same time, that the most important part of proposals for possible changes in the landscape concerns the agricultural lands. In spite of the fact, that the number of these combinations is not big, they occupy almost 49% of the territory of Slovak Republic and they are the most dynamically changing territories.

7.2 Primary Alternative Proposition

This procedure concerns the evaluation and proposition of all other ABC/CLS combinations, which **were not** resolved by limits and preferences in the previous steps. Thus, we evaluate only remaining combinations. In order to avoid errors, formally, we eliminated and resolved a number of combinations, but as we mentioned the remaining combinations take up almost a half of the territory of Slovakia.

The evaluation of ABC/CLS combinations based on selective criteria—current use of ABC types—has the following logic:

(a) If a given ABC (formally a given row of Table 7.1) includes a CLS element—it means such use—which is **identical** with the best sequences of suitable of non-limited activities identified within the results of negative selection (column **Results**), such ABC/CLS combination will be considered positive and the current utilisation of an ABC type will be **maintained**.

This relation can be formally presented in the functional part of the column **CLS—current landscape structure—y_n** in Table 7.1 as follows:

P = code of current activity (i.e. two capital letter codes from the header of the column **Limit and zero values $s^R_{\mathbf{ABC}}$**)

(b) If a given ABC type includes such CLS element, which does not belong to the non-limit activities from the negative selection, the activity is limited in a given ABC type, such ABC/CLS combination will be considered negative and we will propose to change the use with non-limited activity. In Table 7.1, it is presented as

    N = code of a proposed non-limited activity

(c) If a given ABC type includes such CLS element, which **is not limited**, such ABC/CLS combination will be considered as feasible, we can decide to **maintain** or **change** the current use to non-limited activity in accordance with our other criteria. In Table 7.1, it is presented as

    M = code of the current activity, or
    M = code of proposal for a non-limited activity or
    M = code of proposal for a conditionally limited activity

The result of this procedure is determination of one to three most suitable activities—**alternatives**—for each LEC type (Table 7.1). The proposals are inserted into the appropriate cells of the decision table. Consequently, these proposals are projected onto a map.

7.3 Secondary, Variant Proposition—Functional Typisation of Territory

In this stage, each LEC type is assigned an ecologically most suitable use (a function) by means of selection from the alternatives yielded by the primary proposition. It should be noted that this stage applies the criteria of topical and topical-choric indicators of LEC types, but also selected spatial relations of areas concerned. The critical spatial relationships for these proposals include

- size of the area—each activity has demands for topical indicators and size of an area. Special attention should be given to the minimum size of an area;
- shape of an area—activities require certain shapes, and some shapes could even limit certain activities (e.g. narrow areas are not suitable for large-block arable land);
- Vicinity—takes into consideration socio-economic phenomena—protection zones, sanitary zones, aesthetic impacts, perception and other aspects.

These aspects partially exceed the ambit of the LANDEP methodology, and therefore will not be further discussed. However, organisation of spatial relations falls into the group of the methods of basic research, which are intensively dealt with especially with respect to improving the quality of the LANDEP methodology (Miklós et al. 1997; Miklós and Miklisová 1987; Špinerová and Miklós 2007; Moyzeová 2010; Špinerová 2011, Miklós and Špinerová 2013).

The result of secondary proposition is so-called **functional typisation** of territory. It is regarded as micro-organisation of territory.

The functional typisation of territory involves

(a) designation of suitable areas for agricultural activities:

  - suitable for use as arable land of different qualities or various types of rotation of crops.
    Note: Decisions concerning the suitability for specific crops are made only in exceptional cases. Such determination also exceeds the ambit of LANDEP and falls within the competence of specialised institutions.
  - suitable for perennial forage crops, requiring annual ploughing,
  - suitable for different types and intensity of permanent grasslands,
  - suitable for permanent crops—vineyards, orchards, gardens, garden allotments.

(b) designation of areas for activities other than agricultural production, within them:

  - elements of the territorial system of ecological stability TSES
  - areas of landscape greenery to ensure the ecological stability of the landscape (above the biocentres, biocorridors and interactive elements, as anti-erosion, filtration , sanitation, recreation and other functional and aesthetic greenery)

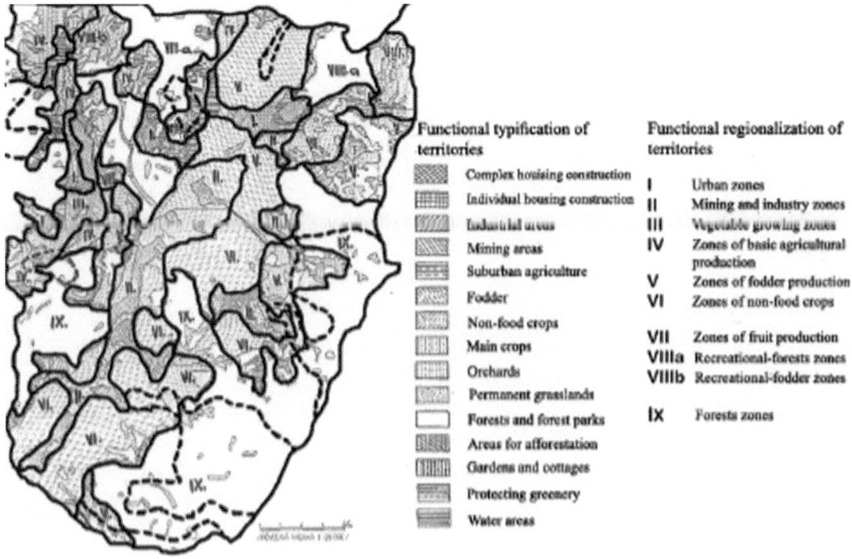

Fig. 7.2 Example of the functional typisation and regionalisation of the territory (region Upper Nitra basin, Slovakia)

- suitable areas for water areas and wetlands performing various functions,
- suitable areas for forests performing various functions.

(c) designation of areas for technical works,

- suitable areas for recreational use,
- suitable areas for residential buildings of all kinds,
- suitable areas for industrial and technical buildings of all kinds.

The proposals determine the most suitable to unsuitable landscape-ecological conditions for performance of individual activities.

This result should serve as a material mainly for physical/spatial/territorial plans of municipal zones and municipalities (see Act No. 50/1976 Coll. on Territorial Planning and Building Code, as amended by Act No. 237/2000 Coll), or for residential conurbations, micro-regions, at the most districts (for example, see Fig. 7.2).

Relation LANDEP Versus TSES

The concept of the territorial system of ecological stability TSES is the most successful landscape-ecological concept in Slovakia legally defined and firmly incorporated to spatial planning processes (Lőw et al. 1984; Buček et al. 1986; Buček and Lacina 1994; Miklós 1994; Izakovičová et al. 2000; Diviaková 2011). As for the relation between LANDEP and of the projects of TSES, it should be emphasised that TSES has been developed on the same methodical base as the LANDEP actually has the same methodical steps (Miklós et al. 2011). TSES is considered an organic part of the ecologically optimal organisation and land use. LANDEP cannot be projected without respecting the elements of TSES as a part of optimal organisation, and also the TSES cannot be established without being a part of ecologically optimal organisation. An ideal case scenario is when the TSES is designed as a part of a LANDEP project. In this case, ecostabilising measures do not need to be projected separately, because a part of LANDEP proposals, which addresses the determination of the functions of agricultural and forest areas and their use acts as a ecostabilising aspect. Where this is not the case and TSES project is provided separately, ecostabilising measures need to be proposed to substitute a comprehensive LANDEP. Failure to do so might compromise the functioning of the TSES and threaten the optimal spatial organisation of the landscape by other projects.

Note: If the TSES has already been projected for an examined territory, it can be accepted as a whole. However, it is necessary to reconsider the harmony of TSES elements and proposals of agricultural and forestry activities and synchronise ecostabilising activities.

7.4 Tertiary Proposition—Functional Regionalisation of Territory

Especially in larger areas, it is possible to observe a distinctive grouping of identical or similar functional types into larger units of regional character. Therefore, this stage aims to determine the most ecologically appropriate function of larger territorial units (zones, subregions, regions, etc.) in terms of their broader territorial (choric) relations. They include, for instance, location, vicinity, distance criteria, size of area, spatial continuity, etc., within a larger area, and their functions are determined by the prevailing use proposed for individual areas (types of landscape-ecological complexes). This is considered as meso- and macro-organisation of territory. The output of this stage is so-called **functional regionalisation of territory**, which expresses proposed general function of a given larger territorial unit in terms of spatial relations. This direction is particularly important at present as it represents a regional approach to the environmental care and constitutes an important element of overall political orientation so-called regional policy.

Typical content of the functional regionalisation of territory is as follows:

(a) At the level of urban conurbation, micro-regions, and districts:

Designation of the principal function based on the prevailing proposed utilisation (functional typing), e.g.:

- residential zones (regions),
- residential-industrial conurbations (regions, zones),
- industrial zones,
- agricultural regions (zones),

- zones of intensive agricultural production on arable land,
- vegetable-growing zones (greenhouse and plastic greenhouse zones),
- pasture-meadow regions (zones),
- recreational regions (zones),
- forest management regions (zones),
- water management regions (zones).

The terms *zones* and *regions* can be used relatively loosely, based on the purpose of an individual landscape-ecological plan (e.g. the term *zone* is used in territorial planning). This direction is regarded as meso- and macro-organisation of territory.

These outputs should serve as the material for land-use plan of a region (see Act No. 50/1976 Coll. on Territorial Planning and Building Code, as amended by Act No. 237/2000 Coll), or associations of municipalities and larger settlements. See the example shown in Fig. 7.2.

(b) At the level of regions, large territorial units, state:

Delimitation of natural residential regions and designation of their functions based on their functional–positional relations, e.g.:

- central (core) regions,
- transit (transitional, linking, crossroads, gates) regions,
- terminal regions,
- peripheral (border) regions, etc.

These procedures are used for ecologisation of planning documents at the highest level, and should serve as a basis especially for Conception of the Spatial Development of the Slovak Republic, but also for landscape planning documents of large territorial units or protected areas and other functionally delineated regions. However, they do not have to be developed within a particular LANDEP project and have broader validity; for specific LANDEP projects, it is advisable to adopt the framework of macro-organisation of territory from other studies.

An example presents the map *Positional functional characteristics of natural-settlement nodal micro-regions* from the Landscape Atlas of Slovak Republic (Miklós 2002). The legend of this map might be of interest to readers (Fig. 7.3).

7.5 Final, Special and Detailed Proposals

As an example of a final, specialised proposal, we present a complex proposal for measures to mitigate the problems of runoff, erosion and accumulation in the catchment area of the Ilijský stream by optimising the organisation and utilisation of the landscape structure. The proposal is based on landscape-ecological analyses and syntheses and subsequent four domains of interpretations and evaluations, namely:

- zone of slope dynamics (see Sect. 5.1.2, Fig. 5.4, 5.5, 5.6),
- ecological quality of the landscape structure (see Sects. 5.2.2, Fig. 5.7),

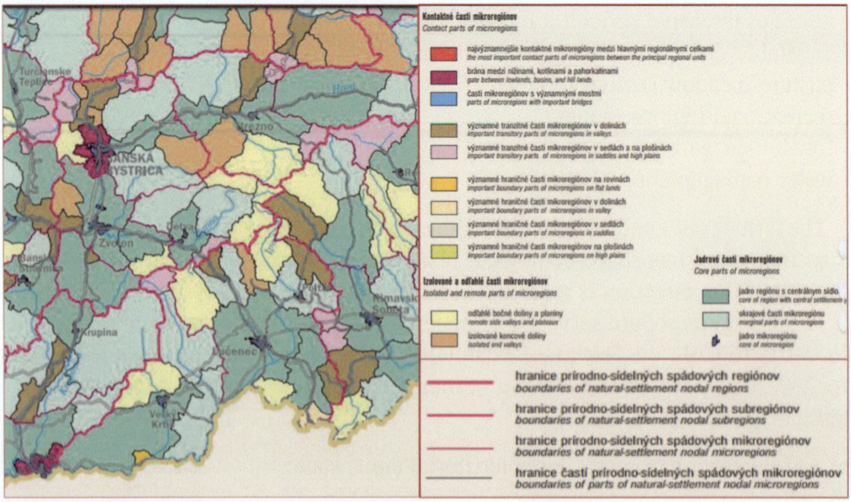

Fig. 7.3 Example of natural settlements regions and their positional functions of Slovak Republic (cut), *source* Landscape Atlas of Slovak Republic, Chapter VII, Map 23 Positional functional characteristics of natural-settlement nodal micro-regions. The original map scale 1:1,500,000. The original legend in Slovak and English

- spatial diversity of the landscape structure by a degree of entropy (see Sect. 5.2.3, Fig. 5.10, 5.11),
- limits of agricultural use of the model territory (see Sect. 6.6.1, Fig. 6.1).

Based on these procedures, it is possible to formulate a proposal for elimination of the threat to the LC types of slope dynamics (Fig. 7.4).

The LANDEP methodology is only rarely applied in its entirety from its analyses up to tertiary propositions, therefore the methods of partial and specialised outputs were developed, which have a wider practical application. However, they are also based on the same landscape-ecological analyses and syntheses, which are interpreted and assessed in various ways. As already mentioned, the most massive partial outputs from the frame of LANDEP are the results of interpretations, as described in the Chap. 5.

The most common examples of partial outputs are also projects of the territorial system of ecological stability TSES that are specifically defined in the legislation. Other possible partial outputs include determination of suitability of territory for hydro-melioration, for the establishment of water areas, ponds, solving the problems of erosion and accumulation, proposal of landscape greenery, proposal of recreational utilisation of the landscape, problems related to the conflict of interests (Izakovičová 2000), spatial projection of the limits of territory for selected activities, problems of land use, problems of flooding, etc.

Fig. 7.4 Example of a combinated proposal to mitigate the problems of runoff, erosion and accumulation (model area of Ilijský creek catchment) (original map scale 1:10,000)

Legend to Fig. 7.4.

Legend key no.	Measures for the optimisation of spatial organisation and utilisation of the territory
11.	Change of the land use: change to grassland Increase the allotment of landscape greenery Plantation of a broad belt of trees and shrubs on the borderline of the zone
12.	Change of land use: increase the allotment of the forage crops area Plantation of a narrow belt of shrubs on the borderline of the zone
21.	Plantation of a narrow belt of shrubs on the borderline of the zone Increase the allotment of landscape greenery Anti-erosion approach to the field cultivation
22.	Plantation of a narrow belt of greenery on the borderline of the zone
31.	Plantation of shrubs to the line of the highest degree of horizontal curvature Increase the allotment of landscape greenery Change the land use: change to grassland or shrubs
32.	Plantation of shrubs to the line of the highest degree of horizontal curvature
33.	Ecologically friendly forest management Anti-erosion measures on the forest roads
41.	Plantation of broad-belt shrubs and trees on the outer borderline of the zones
42	Plantation of broad-belt shrubs on the outer borderline Increase the rate of entropy by landscape greenery
13, 23, 43, 51, 52, 53	Maintenance of the current land use Increase the rate of entropy by landscape greenery

References

Buček A, Lacina J (1994) Mapování biotopů a územní systémy ekologické stability. In: Mapování biotopů, sborník referátů, VŠZ v Brně, pp 59–62
Buček A, Lacina J, Lőw J (1986) Územní systémy ekologické stability krajiny. Životné prostredie 20(2):82–86
Diviaková A (2011) Biotické komplexy pre environmentálny manažment. VKÚ, a. s, Harmanec
Izakovičová Z et al (2000) Metodické pokyny na vypracovanie projektov regionálnych ÚSES a miestnych ÚSES. Združenie krajiny 21, Bratislava
Lőw J et al. (1984) Zásady pro vymezování a navrhování územních systému ekologické stability v územne plánovací projekcii. Agroprojekt, Brno
Miklós L (1994) Koncepcia územného systému ekolgickej stability (ÚSES) na Slovensku. In: Prírodná časť krajiny, jej výskum a návrhy na využitie, zborník referátov. PriF UK KFG, pp 19–24
Miklós L (2002) Positional—functional characteristics of natural-settlement nodal microregions. In: Landscape Atlas of Slovak Republic, Chapter VII, Map 23. MŽP SR Bratislava, SAŽP Banská Bystrica
Miklós L, Krcho J, Hrnčiarová T, Matečný I, Kozová M (1997) Morfometrické ukazovatele reliéfu a ich úcelové interpretácie pre plánovacie procesy. Nadácia Katedry UNESCO pre ekologické vedomie, Banská Štiavnica
Miklós L, Diviaková A, Izakovičová Z (2011) Ekologické siete a územný systém ekologickej stability. Vydavateľstvo TÚ vo Zvolene
Miklós L, Miklisová D (1987) Shape and size of elementary areas and microbasins—evaluation in landscape ecological planning (LANDEP) methodics. Shape and size as spatial categories in landscape ecological planning (LANDEP) methodics. I. part. Bratislava, Ecology (ČSSR), 6(1):85–100
Miklós L, Špinerová A (2013) Krajinné plánovanie. Praktikum. Technická univerzita vo Zvolene, 96 pp. ISBN 978-80-228-2610-5
Moyzeová M (2010) Navrhovanie územných systémov ekologickej stability na vybraných modelových územiach. Životné prostredie 44(3):138–142
Špinerová A (2011) Prvky využitia zeme ako regulátor zón dynamiky svahu v modelovom území Ilijského potoka. Doktorská dizertačná práca. TU Zvolen
Špinerová A, Miklós L (2007) Priestorová diverzita krajiny. Záverečná správa. Katedra UNESCO pre ekologické vedomie a trvalo-udržateľný rozvoj, Banská Štiavnica, Inštitucionálny projekt FEE TU Zvolen č. AE-II 3402 Priestorová diverzita krajiny

Conclusion
Status, Problems and Application of the Results of LANDEP

The output of above-mentioned process are referred to as the ecologically optimal organisation and land use. The ultimate aim of the proposals is delineation and characterisation of integrated territorial units, with unified method of linkage of natural and socio-economic sources, which would represent natural-socio-economic types and regions with optimal internal structure of the use of each area. This principle of functional division of a territory constitutes a basis for ecologisation of regional development and **regional sustainable development**.

According to Chapter 10 of AGENDA 21, comprehensive land-use plans should constitute an **integrating and coordinating** framework for all other sectoral plans and projects and should precede them contentually and chronologically. Despite the general recognition of the importance of a landscape-ecological plan for sustainable development, practical implementation of the proposals for ecologically optimal spatial organization is very difficult. The main reason is that the proposals **require changes** to current land use, which causes many different problems. The principle of spatial organisation of a territory is, however, disrupted by various administrative and economic issues. The methodology of landscape-ecological planning (Ružička and Miklós 1979, 1982a, b, 1990, Miklós and Špinerová 2011), which was developed to address the current state of the spatial planning processes, incorporates all the above-mentioned principles, and its outputs should serve as a basis for spatial/physical/territorial planning, as well as for sectoral planning processes.

The main problems of the application of the optimal spatial organisation can be divided into several groups:

(a) problems with the contents of the outputs and proposals:

Ecological proposals strictly keep to the content and thus the boundaries, shape and size of the types of landscape-ecological complexes, and the resulting form of the proposals reflects the characteristic landscape-ecological structure of a given area. They might respect less the current boundaries and use of plots, ownership relations and other economic aspects of land use. These facts pose a fundamental problem

L. Miklós and A. Špinerová, *Landscape-ecological Planning LANDEP*,
https://doi.org/10.1007/978-3-319-94021-2

arising between the current land use and ecological requirements for changes to the current land use. However, the ecological proposals are based on unchangeable or partially changeable indicators of the primary and secondary landscape structure, therefore it would be wrong to evaluate their utilisation according to whether they agree with the current state, or non-ecological visions of the future state of the landscape.

(b) problems with the recipient and implementer of ecological proposals:

Nowadays, the supra-sectoral content of ideal ecological studies causes that they practically have only few theoretical consumers, who would adequately use all the proposals. This function is not completely fulfilled even by territorial planning which has this defined to produce harmonised and integrated plans for the whole territory and all sectors, let alone by sectoral plans or an individual approach to land use. This causes a conflict of interests and consequent disputes regarding delimitations and recultivations and the logic of the ecological proposals may become disrupted.

(c) problems with the notions concerning the status of ecological materials in spatial plannings:

Although there is a great formal hunger for ecological materials, current notions concerning the importance, functions and status of landscape-ecological plans in practice are highly distorted. They are:

- **minimalistic**—seek to answer the questions such as how much greenery needs to be planted in a district, cadastre or a particular plot, where is it possible to plough etc.
- **maximalistic**—envisage that the development of ecological plans will immediately solve all kinds of problems—water and air will become clean, erosion will stop, and crop yields will increase without initiating any substantial changes to organisation, management and technology.

Nevertheless, it can be claimed that there is a well-founded institutional basis for implementation of landscape-ecological plans in Slovakia. If the ecological principles are to bring a society-wide effect, the methodological status of landscape-ecological plans must be widely accepted: they cannot constitute only a specific material of unclear purpose, but must function as one of the basic documents for each single plan and project that deals with land use, a document equal with technological and economic documents as specified by numerous laws.

Only in that case will LANDEP fulfil the function defined in AGENDA 21 with respect to integrated landscape management, as well as demands of sustainable development defined by current documents including the sustainable development goals in AGENDA 2030 (Transforming … 2015).

References

Miklós L, Špinerová A (2011) Krajinno-ekologické plánovanie LANDEP (The landscape-ecological planning LANDEP). VKU, Harmanec

Ružicka M, Miklós L (1979) Teoretické a metodologické základy biologického plánovania krajiny. Záverecná správa úlohy VI-3-5/1. Bratislava, ÚEBE SAV

Ružička M, Miklós L (1982a) Metodické poznatky ekologického hodnotenia územia pre zónu a sidelný útvar (na priklade Rimavskej Soboty). VEDA, Bratislava, Acta Ecologica, 9, 26, 74 pp

Ružička M, Miklós L (1982b) Landscape-ecological planning (LANDEP) in the process of territorial planning. Ekológia (ČSSR) 1:297–312

Ružička M, Miklós L (1990) Basic premises and methods in landscape-ecological planning and optimisation. In: Zonnenveld IS, Forman RTT (eds) Changing landscapes: an ecological perspectives. Springer, New York, pp 233–260

Transforming our World: The 2030 Agenda for Sustainable Development (2015) United Nations

MIX
Papier aus verantwortungsvollen Quellen
Paper from responsible sources
FSC® C105338

If you have any concerns about our products,
you can contact us on
ProductSafety@springernature.com

In case Publisher is established outside the EU,
the EU authorized representative is:
Springer Nature Customer Service Center GmbH
Europaplatz 3, 69115 Heidelberg, Germany

Printed by Libri Plureos GmbH
in Hamburg, Germany